AF824217

RICARDO REID AND CASEY GERENA

Python 101

Learn the Python Basics in 14 Days

NERD
CHALLENGES

First published by Nerd Challenges LLC 2021

Copyright © 2021 by Ricardo Reid and Casey Gerena

All rights reserved. No part of this publication may be reproduced, stored or transmitted in any form or by any means, electronic, mechanical, photocopying, recording, scanning, or otherwise without written permission from the publisher. It is illegal to copy this book, post it to a website, or distribute it by any other means without permission.

Ricardo Reid and Casey Gerena asserts the moral right to be identified as the author of this work.

Ricardo Reid and Casey Gerena has no responsibility for the persistence or accuracy of URLs for external or third-party Internet Websites referred to in this publication and does not guarantee that any content on such Websites is, or will remain, accurate or appropriate.

Designations used by companies to distinguish their products are often claimed as trademarks. All brand names and product names used in this book and on its cover are trade names, service marks, trademarks and registered trademarks of their respective owners. The publishers and the book are not associated with any product or vendor mentioned in this book. None of the companies referenced within the book have endorsed the book.

First edition

ISBN: 978-1-7374537-1-0

This book was professionally typeset on Reedsy.
Find out more at reedsy.com

To all of our family, friends, co-workers, and the students who encouraged us to write this book, we hope that you enjoy it, but most of all, that this book serves as a stepping stone in achieving your goals.

Contents

Foreword

I am pleased to see how this book turned out, including the authors' refreshing approach with presenting the instructional content in an easily digestible manner, focused on comprehension and retention. Having spent the past 15 years in engineering and technology, both in the industry and academia, I cannot stress enough the importance of understanding technology fundamentals. Python has become a well-recognized programming language, with applications across varying functions and industries. Python has continued to grow in adoption primarily due to the relative ease of learning the language, making it an excellent entry point into programming for technical and non-technical learners.

My focus in the industry has mainly been in professional services, crafting and operationalizing technology investments and transformation initiatives for countries around the globe. A significant portion of my role focuses on preparing strategic, multi-year technology roadmaps and managing large-scale, complex technology programs for client engagements. Our programmers, developers, architects, and engineers have long been the unsung heroes of our technical projects. At the time of this writing, there continues to be a shortage of these roles and associated skill sets in the workforce. Over the past decade, organizations have increased focus on digital transformation as a critical priority, centered on becoming more digital, embedding technology into everyday operations, and upskilling their workforce. I have been on countless planes and sat in many rooms with senior partners and C-suite executives across varying industries, markets, and territories, discussing this shortage along with their plans and roadmaps; the investment in technology and people will continue to grow for years to come.

This book is structured to appeal to individuals on different learning journeys, both technical and non-technical in nature. Whether you are a beginner seeking to initiate your learning journey, an advanced learner seeking to supplement your learning programs, or even someone looking to assess their current knowledge level, following years removed from any development role, this book is well-positioned to aid in your goals. I recommend this book for my teams that I coach, including my project managers seeking to obtain a sense of familiarity in Python and manage their technical product teams. Given the 14-day approach, this book offers the ability to gain value-added insight, regardless of your end goal, in a rapid timeframe.

An apparent characteristic I share with the authors is that we are all lifelong learners. In addition to my role in the industry, I am also an educator, researcher, and peer-reviewed research author. Both authors have held various technical positions in the industry and have continued to hone their craft, engage in learning courses, pursue technical certifications, supplement their functional roles, and keep their minds sharp. With that said, I am pleased with the approach the authors took with crafting this book; they leveraged the best practices from their professional and academic experiences, and it shows – this book is a lean, straight-to-the-point approach, providing the essentials for learning Python, with no "fluff" and only value-added content. An added differentiator is their unique "Nerd Challenges" strategically sprinkled throughout the text to provide engaging, hands-on learning. Any educational text that gets readers out of simply reading, and encourages fingertips on the keyboard for real-time practice, clearly understands how to facilitate (and accelerate) learning. The authors understand and employ this approach masterfully in this book.

> *Josh Lumseyfai, Ph.D., is a trusted, technical professional with a continuous improvement mindset, as well as a published, peer-reviewed research author. Dr. Lumseyfai is also an Adjunct Professor at an accredited private institution. His experience falls within the inter-disciplinary fields of engineering, technology, consulting, program management, and strategy. Dr. Lumseyfai currently serves as a trusted strategic technology advisor for a global professional services network, crafting and operationalizing technology investments and transformation initiatives for over 150 countries around the globe. He is recognized for his strategic, collaborative, and "systems thinking" approach. Dr. Lumseyfai has excelled at leading large-scale, complex programs, coaching high-performing teams, and aligning business and technology transformation initiatives. Dr. Lumseyfai holds several professional certifications, including with Amazon Web Services (AWS), the American Society for*

Engineering Management (ASEM), the American Society for Quality (ASQ), Google, Human Factors International (HFI), IBM, Microsoft, PricewaterhouseCoopers (PwC), the Project Management Institute (PMI), and Salesforce.

Preface

Before you jump into reading the book, Casey and I just wanted to point out a couple of things that will be helpful to you as you begin your Python journey. Below is a list of tips, resources, where to find solution code, and the book's schedule.

Is this the right book for you?

Here are some questions to ask yourself in order to determine if this book is right for you.

- Are you a beginner programmer?
- Are you interested in learning the fundamentals of Python?

If you answered yes to both of those questions, this is the book for you. In this book, Casey and I solely cover the fundamentals of Python. If you are looking for more advanced topic matter related to Python there are better suited books and courses available to you. I've listed some in the **Other resources** section.

Solution code

You can find all of the solution code for each day's content in our GitHub repository, which is located here: https://github.com/nerdchallenges/python-101-book

Book Schedule

It's essential to stick to the book's outlined schedule; the reason for this is simple. Casey and I have found over the years of teaching junior software engineers and people new to programming that it's effortless to feel overwhelmed and get lost in what you don't know. It's also become increasingly difficult for many people to stay on track. Casey and I have outlined this book so that you can cover a few key topics each day, recap the activities from that day, all in about 20 minutes of reading.

In other words, you should be able to finish one day's worth of material during your lunch break. We know what it's like to pick up a 1,000-page software engineering book and just feel exhausted before you make it past page three. We wrote this book to help address that. So stick to the outline, and you'll be fine. If you'd like to cover two days worth of material in one day, go for it. However, I would not recommend moving more than two days at a time; it's good to give yourself time in between sessions, so you have time to think and process what you just learned about.

Reading Time

At the beginning of each day you'll find an estimated reading time, it isn't a number that is set in stone it just gives you an idea of how much time you should set aside to read the entire day. Some days it may take you longer, some days shorter, it's just general guidance. On average each day should take roughly 20 minutes; however, there are some exceptions, so plan accordingly!

Code Snippets and Important Terms

Unfortunately, in terms of formatting a book you are rather limited in terms of what can be displayed, especially if you want to support eReaders. As such, you'll find that most of this book is in black and white (so no colors). It's important in programming to be able to determine what is a piece of code, what is a comment (more on this later), and what is output from your terminal. Below are small examples of what each of those things should look like.

Code Example

```
print('Hello World!')
```

Code Example with Comments

```
# This is a comment
print('This is some real code!')
```

Terminal Output

Terminal output will be described as a *Result* and will appear beneath a code block, like this

```
answer = 2 + 2
print(answer)
```

Result: 4

Other resources

- **Solution code:** Can be found in our github.com repo - https://github.com/nerdchallenges/python-101-book
- **Official Python site:** The official place of all things Python - https://www.python.org/
- **Udemy course:** If you prefer to watch videos as opposed to reading or you want to supplement this book with lecture material, the course is offered on Udemy - https://udemy.com
- **Nerd Challenges:** Our official site for other books about Technology and Fitness - https://nerdchallenges.com
- **High Performance Python:** Excellent resource on optimizing Python code, for a more advanced engineer - https://www.amazon.com/High-Performance-Python-Performant-Programming/dp/1492055026/

Acknowledgement

Thanks to Ivan Lozano and Sean Soto for assisting us with technical reviews of the book's content, your consistent feedback throughout the book was instrumental to completing it on time.

Errata / Contacting Us

We're big believers in continuously growing and improving. If you'd like to share your thoughts or give us any feedback, we would love to hear from you. We write our books for our readers, so if something isn't working for you or you want us to keep up the excellent work, please feel free to reach out to us at rico@nerdchallenges.com or casey@nerdchallenges.com.

Feedback and Reviews

We wrote this book because we wanted to help anyone who was interested in Python programming but might not have any prior experience with programming. So, if you found this book beneficial or if you didn't please leave us a review wherever you purchased this book from. We do our best to read every review and in many cases respond to the review (if possible), so your feedback is critical and extremely important to us.

Introduction

Have you ever thought that programming was way too challenging to learn? Perhaps you've said the all too common phrase, "I'm just not technical enough." We've heard these concerns and more from our students, and based on our experience, they aren't true. We believe anyone can learn to program regardless of their educational background or technical expertise. No degree or special training is required to learn a programming language, and becoming a programmer is not something that takes years of training.

In this book, we teach you the basics of programming in Python in only 14 days with a simple investment on your part of only 30 minutes to an hour each day. We use a challenge-based approach which we believe accelerates learning by incorporating milestone projects, or what we like to call "Nerd Challenges", throughout the book to get you hands-on experience with the concepts you are learning. We find that this reinforces the concepts we teach and increases retention in our students, resulting in a more efficient and enjoyable learning process.

What is Python?

Python is an interpreted high-level programming language. It was created by Guido van Rossum in the 1980s and initially released in February of 1991. The code structure stresses a heavy emphasis on code readability. Thus, the reason it's such an easy programming language to learn. It is easy to understand and easy to read and write. Python is ideal for anyone learning how to code in a programming language.

Since it is an **interpreted language**, Python code or instructions are executed immediately by a computer program called an interpreter. There is no need to compile the code into a machine language program similar to most other programming languages.

There are other programming languages out there. Some of them operate in the same manner as Python. Main differences in the languages usually stem from how the code executes and the syntax. I stated earlier that Python is an interpreted language. Some languages utilize compilers to run code. Compilers translate source code from the programming language into machine code to be executed by the **CPU (Central Processing Unit)**. Some examples of compiled languages are C, C++, Go, and Rust.

Each programming language also has a **syntax**. The syntax is a broad set of structured rules, processes, and principles that defines correct code. In this book, you'll learn Python syntax. In my opinion, this syntax is the easiest of any language. The great thing about Python IDEs, which I will discuss later, is that they alert you of any syntax errors in real-time with the code you are writing. You can fix incorrect code by determining a syntax error quickly.

Other commonly used languages:

- Java
- C#
- PHP
- Ruby
- JavaScript

Why Python?

Python is easy to write, and it's straightforward to read and even easier to understand. Why else wouldn't you want to learn Python? Well, what does that mean, and how can you use it? Yes, it's a programming language, but used for what? Like anything, you have advantages and disadvantages while using this language. Let us look into some of these before jumping into Python.

Advantages

There are many advantages to using Python, but there are four main advantages: Free and open-source, extensive libraries, works on many platforms or Operating Systems, and dynamic typing. Open-source means a community filled with free libraries shared by other Python developers. You can use code that others have already developed into your projects and code.

The bottom line is that there is no need to reinvent the wheel while programming. Code and libraries have already been created for you that you can use. That leads us to our next benefit over other languages. There is an extensive number of libraries available to be used. I will go over libraries, and you will see that they can make developing our programs easy as 1,2,3!

You can use Python across many operating systems. The advantage of cross-platform development is that it gives you the ability to code an application once and run that same code on different operating systems such as macOS and Windows.

When declaring variables, you have to identify a variable data type in most languages. This strict declaration can make programming difficult and cause errors if set improperly. Fortunately, Python doesn't have this issue. Python is a **dynamically typed** language. When you create a variable, you don't need to tell Python what type of variable it is. The dynamically typed characteristic is one of the many reasons it's so easy to write in Python.

Python can work with almost all databases that you may know. It has a software library called PANDAS that allows data manipulation and analysis in Machine Learning applications. You can import data from various file formats such as SQL, JSON, Microsoft Excel, and CSV files.

Disadvantages

Like with anything, there are disadvantages. I will show you those now to ensure you choose to use Python for the correct applications. This language runs at a slower speed than many other languages. It is a direct result of being an interpreted language rather than a compiled language. Essentially the code is being executed much slower because it's converted to machine code at run-time.

The interpreter, while helpful, does pose other issues of its own. It also performs memory management of the program.

You do not control how Python allocates memory while the program is running. Memory management is not a showstopper in the smaller programs you will be dealing with in this book. However, it is good to keep that in mind when designing more extensive programs.

In compiled languages, most of your errors will occur at build-time. The compiler must compile or build your code for the CPU to run it, and if any errors occur, the compiler will notify you.

In interpreted languages, errors occur at run-time because there is no build or compiling step. These run-time errors arise during the program's execution and can render the program entirely inoperable. You will be exposed to the common run-time errors so you can avoid them while creating programs.

When not to use Python

All in all, Python is a great language; however, it isn't appropriate for every single use case. There are a few key areas where it doesn't shine. First, if **speed** is a significant concern for your application, Python, or any interpreted language for that matter probably shouldn't be used. Compiled languages just run fast, period. Second, **mobile application** development isn't generally Python's strong suit. While there are newer frameworks such as Kivy (https://kivy.org) and Beeware (https://beeware.org), you're limited in what you can build. You're better off sticking with the native languages Swift (iOS - https://developer.apple.com/swift/) and Kotlin (Android - https://kotlinlang.org/). Third, generally speaking, any application that requires a Graphical User Interface (GUI). Python doesn't have as many rich GUI or graphic libraries to choose from that are well supported. As you'll see in one of the Nerd Challenges, you'll build a Python-based GUI calculator program. It's cool; however, it just doesn't look all that appealing. So if you're making a 3D game, Python probably isn't the best choice. Finally, Python isn't well suited to build low-level applications such as Operating or Embedded systems.

What can you build with Python?

The reason you are here is to learn Python. I discussed the advantages and disadvantages of using Python. You should be asking yourself now is what can it be used for and how? The applications for Python are endless. What if you had a single repetitive task that you wanted to run over and over? You could write a Python script. That script would allow you to efficiently run a program to perform that single task as many times as required. It makes running a set of predefined instructions as simple as pressing the run button on your schedule.

Maybe you have 20 things on your computer that you do every day. It might take you 2-3 hours per day to get through accomplishing those 20 things. Why not create a script that can perform those 20 things with the touch of a button and execute those activities quicker in about 5% of the time it would take you to do it manually? While developing software, there may also be a need to test functionality. Scripting can save time by creating standardized test scripts. From the example, you can see how scripting leads to automation making our tasks more straightforward and efficient. Some other Python use cases are:

- Quickly create web applications and sites with Python frameworks. Frameworks such as Django and Flask make creating a website relatively easy in less time.
- Design single desktop applications capable of running across all Operating Systems (e.g., Microsoft Windows, macOS, and Linux distributions).
- Analyze large amounts of data to extract information to make more informed decisions.
- Use libraries and frameworks for Machine Learning and Artificial Intelligence projects, such as PyTorch (https://pytorch.org/).
- Simplicity and readability make it the primary choice to learn programming concepts. Python can act as a gateway into learning other high-level programming languages. Many college universities today across the U.S. and the world have students learn Python before other languages.

Hopefully, that gave you a good understanding of all types of things Python can and cannot do. It's time to stop talking *about* Python and time to start programming it! Remember, you only have 14 days to learn to program/code in Python. Let's get started!

1

Day 1: Installing Python, Variables and Comments

Reading Time: 23 minutes

Today, you're going to dive straight into setting up your environment to start coding in Python. You are installing the interpreter, which I discussed in the Introduction. You are also installing an Integrated Development Environment or IDE. The IDE gives you a friendly user interface that provides valuable tools to develop your programs. After that, you'll be learning about variables and how to place comments in a Python application.

Throughout this book, you will be using Python 3. Remember that Python 2 is considered the legacy version and is still used by some applications. Python 3 is similar to Python 2, so if you know Python 3, you know Python 2. It won't be challenging to understand a program written with the legacy version. Python 3.10.0 is the latest version; however, know that all of the code you will be writing in this book can be used with version 3.7 or higher. In the sections that follow, you will be installing two primary things, Python and the IDE, PyCharm. In each section, you'll find instructions for each operating system, so for each tool, skip to your OS and follow that portion only. One exception to this is for macOS. You'll be installing an additional tool called Homebrew, which you'll learn about in that section.

* * *

Python Installation

Installing Python on Windows

I'll be going over how to install Python on Windows; however, regardless of the Operating System you have, you should be able to successfully install Python by following the instructions on the Python documentation page, located here https://wiki.python.org/moin/BeginnersGuide/Download

Identify the Operating System (OS) you are currently on and the system type. The system type indicates whether the OS is 32-bit or 64-bit. Most modern laptops are 64-bit; if you have an older system, perhaps choose 32-bit. Once you have identified the OS, open your web browser and navigate to https://www.python.org/downloads/, the official download page for Python. The page should open up, showing the latest version for your Operating System. In my case, the newest version at this time is Python 3.9.5. Click on Download to download the latest stable version of Python.

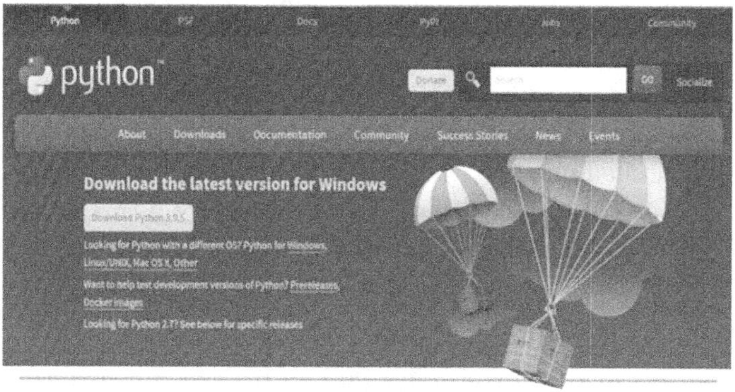

On Windows, this downloads an executable. Go ahead and open it up and accept all of the details. Select **Add Python 3.9 to PATH**.

Python, IDLE, pip, and documentation are installed on your computer when you finish the installation.

- IDLE is an integrated development learning environment bundled into the Python distribution. It is a shell that we're able to write python code. Think of it sort of like a command prompt.
- *pip* is a Python package management system used to manage and install software packages.
- Documentation includes Python how-to guides and a tutorial.

Installing Python on macOS

If you have a newer Mac, likely, Python 3 is already installed. To double-check type in the following in your terminal window.

```
python3 --version
```

You should see something similar to Python 3.8.9. Anything greater than Python 3.7 should suffice for the content in this book.

If you are running on an older Mac and don't have Python 3 installed, you'll want to install it. The easiest way to do this is with a tool called Homebrew.

Installing Homebrew (If Python3 is not installed or if you just want to run the latest and greatest)

Homebrew is a developer tool that allows you to easily install packages and software tools on macOS and Linux. Installing it is pretty simple, open up a terminal and run the following command:

```
/bin/bash -c \
"$(curl -fsSL \
https://raw.githubusercontent.com/Homebrew/install/
HEAD/install.sh)"
```

Then simply follow the instructions and accept the defaults. You should see something similar to the following output. Note: It may prompt you for your admin password. This is fine.

```
==> This script will install:
/opt/homebrew/bin/brew
/opt/homebrew/share/doc/homebrew
/opt/homebrew/share/man/man1/brew.1
/opt/homebrew/share/zsh/site-functions/_brew
/opt/homebrew/etc/bash_completion.d/brew
/opt/homebrew
==> The following new directories will be created:
/opt/homebrew/bin
/opt/homebrew/etc
/opt/homebrew/include
/opt/homebrew/lib
/opt/homebrew/sbin
/opt/homebrew/share
/opt/homebrew/var
/opt/homebrew/opt
/opt/homebrew/share/zsh
/opt/homebrew/share/zsh/site-functions
/opt/homebrew/var/homebrew
/opt/homebrew/var/homebrew/linked
/opt/homebrew/Cellar
/opt/homebrew/Caskroom
/opt/homebrew/Frameworks
==> The Xcode Command Line Tools will be installed.

Press RETURN to continue or any other key to abort:
```

Press RETURN and let it install everything it needs. It will probably take several minutes for Homebrew to download and install everything it needs, so grab some brew ☕

Once Homebrew is installed, you should see a next steps section with two commands. Copy both of those commands and run them. It should look similar to the below but not precisely! So don't copy these. Copy the ones in your terminal.

```
echo 'eval "$(/opt/homebrew/bin/brew shellenv)"' >>
/Users/administrator/.zprofile
eval "$(/opt/homebrew/bin/brew shellenv)"
```

Now you can install Python by typing:

```
brew install python
```

Tada, all done!

Installing Python on Linux (Specifically Ubuntu 20.04)

Ubuntu 20.04 comes with Python 3 installed, so you should be good to go right out of the gate! Linux is awesome! As my Father always used to say though,

> *Trust, but verify.*

So how do you verify? Run the following command in a terminal window:

```
python3 --version
```

If you see something like the following, you're good to go!
Result: 3.8.10

But wait...This is Python 3.8.10. What if you want the latest and greatest version of Python? You'll then need to compile from source. Compiling from source is a bit beyond the scope of this book. However, it isn't too complicated. It just requires a few additional tools. If you'd like to go down that path, here is an excellent document showing you how to do so https://devguide.python.org/setup/

If you prefer not to compile source, don't worry. Everything we'll be doing in this book will work perfectly fine on Python 3.8.10

PyCharm Installation

You could always write code in a text editor and then run that text file with the extension .py in a command line. That doesn't seem appealing for anyone just starting to learn to program. In this book, you use a Python IDE instead, making writing and understanding code manageable. There are multiple IDEs to choose from, but in this book, I use PyCharm. PyCharm is an application developed by the company JetBrains.

If you want to use a different IDE, it's no problem! Just know that some things, mainly Day 11, on debugging, are explicitly written for PyCharm.

Installing PyCharm on Windows

Navigate to this URL, https://www.jetbrains.com/pycharm/download/ to download the **Community (Free)** version of PyCharm, which is open-source. The professional version is more feature-rich; however, you do not need it for this book's purposes. You can download the file, open it and follow the setup instructions to finish installing PyCharm.

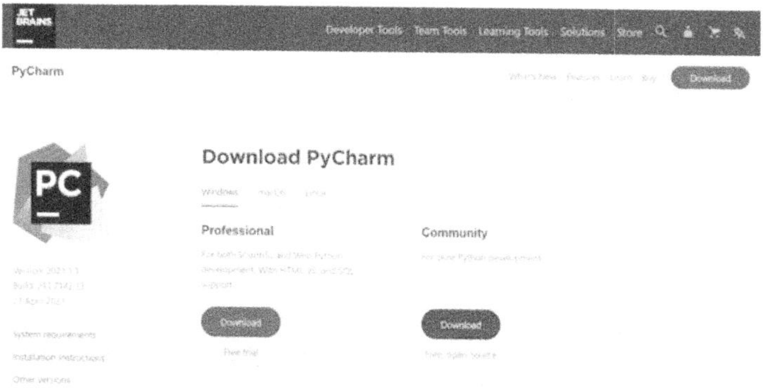

Installing PyCharm on macOS

Navigate to this URL, https://www.jetbrains.com/pycharm/download/#section=mac , to download the **Community (Free)** version of PyCharm, which is open-source. The professional version is more feature-rich; however, you do not need it for this book's purposes. You can download the file, open it and follow the setup instructions to finish installing PyCharm.

> **Note: Make sure you choose Apple Silicon if you're using the latest M1 Chip**

Installing PyCharm on Linux

Navigate to this URL, https://www.jetbrains.com/pycharm/download/#section=linux to download the **Community (Free)** version of PyCharm, which is open-source. The professional version is more feature-rich; however, you do not need it for this book's purposes. You can download the file, open it and follow the setup instructions to finish installing PyCharm.

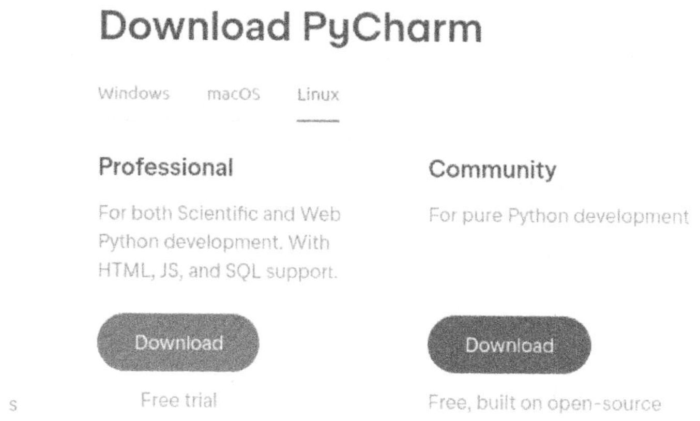

A Note about IDEs

An IDE is very much a personal choice, and there are many suitable IDEs available. I've listed some here for your convenience. Feel free to look at them and try them all out for yourself.

- Visual Studio Code (VSCode): An excellent IDE, which I love. - https://code.visualstudio.com/
- PyCharm: It's incredible - https://www.jetbrains.com/pycharm/
- Sublime text: Another great editor - https://www.sublimetext.com/
- Notepad: Not really an IDE, but I just want to point out that you can use it!

> Important Note: I will discuss a few PyCharm features that I find very useful in the book. Other IDEs may have these options available, but they may not work quite the same way. You'll be on your own to explore those options.

Hello World!

After installation, you're ready to use PyCharm. Open the PyCharm Community Edition application. The Create Project window should appear, as seen below. Go to file at the top of the window, then select **New Project**...

Enter the project's name as **PythonNerd** instead of the default text **pythonProject** in the location field. All other default settings within the Create Project window are good to go. Click Create to continue. Your new project environment should take a few minutes to load.

Now that you have your project built let's look around the workspace. On the left-hand side of the screen, there is a list of your project files and folders. Below that, you'll also notice folders called libs which are internal and external libraries. I will talk more about these later in the book. The file with the name main.py is a python file. All python files or scripts/programs need to have the extension .py to run.

The interior window to the right is where you place your Python code for each script. The window currently displays the code for your main.py script. Select the green triangle or run button in the top right corner to see the script in action. A window directly below your workspace should magically appear with the execution of the main.py code. Congrats on running your first Python script!

Result: Hi, PyCharm

* * *

Introduction to Variables in Python

Here you are focused on some of the basics of Python. I'll introduce you to variables, a few data types, and how to print statements out to view program outputs, as you saw in the program at the end of the last section.

Make sure your PyCharm PythonNerd program window is open and ready for us to use—Right-click on your Program folder off to the left and select New. In the menu, choose **Python File**. A menu pops up, prompting for a name.

Name the file **intro_to_variables.py**. Notice the filename is all lowercase and has underscores instead of spaces. Using underscores to represent spaces is standard practice in Python. The Python files you are creating are scripts or programs. There are other Python files that you will learn about called modules, packages, or libraries. They all use the .py extension!

When writing code, it's always ideal to follow best practices or widely known standards. Fortunately, Python has something known as the Python's Developer's Guide. This guide has documentation known as Python Enhancement Proposals (PEP), proposals that define coding standards for Python. In this book, I will be using PEP 8 entitled *Style Guide for Python Code* standard.

The guide states, "One of Guido's key insights is that code is read much more often than it is written. The guidelines provided here are intended to improve the readability of code and make it consistent across the wide spectrum of Python code." We must keep this legacy going by upholding these standards. Time to dive in!

Understanding Variables

One of the essential elements of a program is the use of variables. A **variable** is a name that holds a value in memory that can be accessed later on in a program. That sounds a little complicated. Let's illustrate this with an example.

Say I want to give a user that's also a nerd named Casey. I can store the name **Casey** in a variable called **nerd_name** with the following code:

```
nerd_name = "Casey"
```

Wow, that seemed effortless. I just assigned the name Casey to the variable **nerd_name**. I can now use the variable **nerd_name** to represent a nerd named Casey throughout the program.

Notice that there are quotations marks around "Casey." When creating variables of text in Python, these are required. Text assigned to variables is known as **Strings**. The name Casey in our example is a string. Suppose I changed the name of **nerd_name** to **"Kelvin"** The text Kelvin is now a string that the rest of the program can access by referencing the variable.

```
nerd_name = "Kelvin"
```

Leaving out a quotation could result in an error while running the program. It can also lead to the variable being inaccessible to the rest of the program. IDEs make it impossible to leave out one of the quotes. You'll get a warning right away if you forget. IDEs are great for double-checking our work on the fly!

Variables can store much more than just strings. They can store other data types such as numeric values, tables, lists, tuples, sets, and dictionaries. More to come on all of these data types as we proceed with the book.

Naming variables

Correctly naming our variables is essential! It is necessary to keep it simple and descriptive enough to indicate what that variable is storing. Say, for instance, you wanted to store a student name in a variable. You might name that variable **student_name** or **studentName**. If you wanted to reserve that student's age, you might call the variable **student_age** or **studentAge**. Please note that all variable names must be unique. If you have more than one student, you could assign a number such as **student_name3** or **studentAge3** for the third student variable.

You may have noticed a few things with these variable names you created. Each variable starts with a lowercase letter. None of our variables should begin with a capital letter or any special characters. You cannot use spaces for variable names. In this book, I use the format with underscores to name variables. So, each variable follows the **student_name** format. Using underscores is a preference of mine and the PEP 8 style guide. So, technically you can choose whatever you'd like. However, I would strongly encourage you to use the PEP 8 style, as it's what you'll most likely see in other Python projects.

Data Types

Strings

You learned about strings already. In our example, my string was just one word. What if you wanted that variable to contain more words? Is that possible? Yes, a string variable can contain more than one word. You can have entire sentences assigned to a string variable. Let's see this in action by setting a sentence to the variable called **nerd_welcome**:

```
nerd_welcome = "Welcome to the Nerd Challenges
Python 101 Course."
```

The assignment above sets the string **"Welcome to Nerd Challenges Python 101 Course"** to the variable **nerd_welcome**.

Numbers

In addition to string datatypes, there are also number datatypes. You can save numbers directly into variables in the same manner. The only difference is that you would leave off the quotations. Let's give our nerd the age of 34:

```
nerd_age = 34
```

Since I left off the quotations, Python assigned the variable as a number. The **nerd_age** variable would still work if quotes were left on, but we would save the age as a string. Storing numbers as strings is a common mistake that can lead to programming errors when conducting mathematical calculations. When you want to assign numbers to variables, leave off the quotation marks.

Now that was reasonably simple. Let us talk more about numbers. The number 34 is a whole number. In Python, a whole number is called an **integer**. Ok, but what is a decimal number called then? A decimal number is known as a **float**. The difference between an integer and a float variable may not seem significant now but will be in later lessons.

Mathematical Operators

With numbers, assignments also come the ability to assign math operations. If, for some reason, that nerd had a birthday today, I could increment his age by adding the number 1, like so.

```
nerd_age = 34 + 1
print(nerd_age)
```

Result: 35

The **nerd_age** variable is now assigned a value of 35. That is right, and Python treats **34 + 1** as a mathematical expression. The plus sign is one of many arithmetic operators available for use in Python. The table below shows some of the standard operators for primary math assignments.

Operator	Name	Example
+	Addition	x + y
-	Subtraction	x - y
*	Multiplication	x * y
**	Exponent	x ** y
/	Division	x / y
%	Modulus	x % y

You are likely familiar with most of these operators except for the last one called **Modulus**. Modulus returns the remainder of dividing x by y. Let us use the previous **nerd_age** variable to illustrate this. For values x and y, I used the numbers three and two. So x = 3 and y = 2 respectively.

```
nerd_age = 3 % 2
print(nerd_age)
```

Result: 1

By setting **nerd_age = 3 % 2**, I assigned one to the variable nerd_age. One is the remainder you receive after dividing three by two. I can continue to do this with any two numbers to return the remainder.

** denotes the exponent operator. Again, I used the variable nerd_age with numbers three and two. Three raised to the second power is nine. The value assigned to the **nerd_age** is now **9**.

```
nerd_age = 3**2
print(nerd_age)
```

Result: 9

I hope you enjoyed this mathematics lesson in programming. I almost forgot that this was a book to teach us how to program!

Print Statements

You have learned how to assign numerical values and strings to variables. In some cases, you even learned how to perform math operations as assignments. There were no visual outputs from running any code I wrote to assign variables. Thus lies the most crucial question: How do you know that your code returns the expected values? Think about it for a second, what if you cannot calculate two raised to the 20th power in your head? How do you check something like that and ensure the code is correct? The answer to this question is the **Print Function**.

The print function is an essential function you'll learn in programming. The print function allows you to output or display values once you execute or run the program. Remember when you ran the main.py program, it displayed Hi, PyCharm.

That was the direct result of using the print function and running the program. You use the following syntax print(string) to use the print function.

Head back to your IDE to view your **intro_to_variables.py** script. In the right window pane, input:

```
print("Welcome to the Nerd Challenges Python101
Course")
```

Result: Welcome to the Nerd Challenges Python101 Course

Don't forget those quotation marks! Right-click the script in the Project files. Then select the green **Run** button to run the **intro_to_variables.py** script.

If you followed my instructions to a tee, you would see the output above. Congratulations on just writing and running your first-ever Python program! You have taken the first leap at becoming a programmer in Python.

Printing out that string with the Print function was pretty cool, huh? Why stop there when you can keep printing stuff out. That was a simple string. What if you decided to print that welcome statement out multiple times? You could do this.

```
print("Welcome to the Nerd Challenges Python101
Course")
print("Welcome to the Nerd Challenges Python101
Course")
print("Welcome to the Nerd Challenges Python101
Course")
```

Result: Welcome to the Nerd Challenges Python101 Course.

Welcome to the Nerd Challenges Python101 Course.

Welcome to the Nerd Challenges Python101 Course.

That seems to work pretty decently. Try to print this out yourself. It works but does not appear to be too efficient. Suppose you had to print an entire paragraph in different locations over and over within your program. In that case, this could quickly become a real hassle. Luckily for you, earlier, we learned how to use variables.

You learned about the variable **nerd_welcome** at the beginning of today, remember? You set **nerd_welcome = "Welcome to the Nerd Challenges Python101 Course"**. So, suppose you wanted to print out the previous example without rewriting the entire string over three times. In that case, you could incorporate the variable instead into the print statement. See how you could simplify the last example with the code below.

```
nerd_welcome = "Welcome to the Nerd Challenges
Python101 Course."
print(nerd_welcome)
print(nerd_welcome)
print(nerd_welcome)
```

Result: Welcome to the Nerd Challenges Python101 Course.

Welcome to the Nerd Challenges Python101 Course.

Welcome to the Nerd Challenges Python101 Course.

The code above yields the same results as before, with much less writing on our part. First, I assigned our statement to the variable. Then, I printed out that variable using the print function three times. Notice that you remove the quotes between the parentheses when you print variables.

Another pertinent concept to grasp is that indentations in Python make a difference. You need to ensure proper spacing and indentations, in this case, are the same. If you indent any of these lines, you could end up with an error while trying to execute the program. I'll talk about indentations more as we discuss functions on a later day.

Single vs. Double Quotes

I have been writing print statements with double-quotes. There is also the option to use single quotes to accomplish the same thing. So, why the need for single quotes? The perfect example would be if someone wanted to print out a statement that already had quotation marks.

For example, let's say I want to output

Casey says, "Welcome to the Nerd Challenges Python 101 Course."

It seems easy, but I cannot just use

```
print("Casey says "Welcome to the Nerd Challenges
Python 101 Course."").
```

If you run this code, Python will return a **SyntaxError**. The quotations won't get you the output you want. You must use single quotes to print statements that have quotes. Let's correct our code, so it now outputs what you are looking for:

```
print('Casey says "Welcome to the Nerd Challenges
Python 101 Course"')
```

Result: Casey says "Welcome to the Nerd Challenges Python 101 Course"

Escapes

You can use **escapes** to print out the same statement. They allow you to insert characters that are usually not valid within a string. You use a backslash followed by the character you want to display to insert an escape. For example, if you wanted to print the exact string as you did previously, but using an escape, you could do the following:

```
print("Casey says \"Welcome to the Nerd Challenges
Python 101 Course\"")
```

Result: Casey says "Welcome to the Nerd Challenges Python 101 Course"

Code	Result	Example
\n	New Line	"Hello Nerd"
\r	Carriage Return	"Hello \r Nerd
\t	Tab	"Hello \t Nerd"
\\	Backslash	"\\"
\'	Single Quote	"'"
\"	Double Quote	" \" "

Concatenation

Concatenation in Python means joining strings together using the + operator. How do you combine those strings to print out one statement if you have multiple different strings? In this example, you have the following three strings:

1. Hello
2. Casey

3. Welcome to Nerd Challenges Python 101 Course!

Let's concatenate or combine these strings to provide a single print statement. You start by using the print function. In the parenthesis, put each string and separate them by + operators. Look at the code below for a better illustration, and try running it yourself!

```
print("Hello " + "Casey" + " Welcome to Nerd
Challenges Python 101 Course!")
```

Result: Hello Casey Welcome to Nerd Challenges Python 101 Course!

You have appropriately spaced out each string by adding spaces inside the quotes. It's good practice to leave a space after each string when using concatenation.

Well, that was easy enough. Let us move on to concatenation with strings and variables. Clear your workspace from any code. You are going to print out the same statement this time with using a variable for the name as follows:

```
nerd_name = "Casey"
print("Hello " + nerd_name + " welcome to Nerd
Challenges Python 101 Course")
```

Result: Hello Casey welcome to Nerd Challenges Python 101 Course

Look at that! Your program returned the same result! The only thing to keep in mind here is to add an extra space at the beginning of the last string to keep spacing. The variable, **nerd_name**, does not have any spaces included; you would not need spaces in the print statement if it did.

String Formatting - Modern Style format()

It's easy to combine strings and variables when the variable is also a string. If the variable is a number, however, this changes everything. Our previous method of concatenation doesn't work with numbers. Remember, concatenation is only for joining strings. There is, however, another way we can combine a string and number variable. For this, we can use another function called *format()*. The format function allows us to insert numbers into strings.

Ok, now let us try to print out the statement "We are on Day 1 of the Python 101 book. I am 34 with 0 years of programming experience," using a combination of strings and variables. Initially, I'm going to create variables for all the numbers. Then I need to create a variable for the string to be printed out by substituting the numbers in the string with open {} brackets. Clear your workspace and paste the code below:

```
book_day = 1
python_series = 101
age = 34
programming_experience = 0
statement = "We are on Day {} of the Python {} book.
I am {} with {} python experience."
print(statement.format(book_day, python_series, age,
programming_experience))
```

Result: We are on Day 1 of the Python 101 book. I am 34 with 0 python experience.

The variable in the format function corresponds to its position in the string statement. For this example, the variable book_day is substituted for the first {}, python_series for the next {}, age for the next {}, and finally programming_experience for the last {}. There is no limit to how many numerical variables you can include.

String Formatting - Legacy % Operator

The % operator was previously used earlier to perform arithmetic operations. The % operator can also format a fixed set of enclosed variables in a list. It's not a commonly used practice anymore. Still, it allows you to quickly identify strings using the % operator in legacy code or Python 2.6 or earlier. The only way to quickly explain how the % operator works is to see it in action.

If you wanted an output that shows
"Hello Nerd Challenges LLC"
we might write the code

```
print("Hello Nerd Challenges LLC")
```

To write this using the legacy % operator, you could substitute the company name with %s, which indicates a string variable is inserted at this location. You would then write **%** with the **company_name** at the end of the statement. Let's jump right into this.

Clear your workspace from any previous code you have. Now write **"Hello Nerd Challenges LLC"** using the legacy code format below:

```
company_name = "Nerd Challenges LLC."
print("Hello %s" % company_name)
```

Result: Hello Nerd Challenges LLC.

The code above yields the same output as just a print statement while utilizing the legacy format operator. Note that you used *%s* because our variable **company_name** was a string, and you wanted to print out a string. What if you wanted to print out a decimal variable but output it as a whole number or integer? That is simple. You can show this with someone's age. See the example below:

```
nerd_age = 34.5
print("This user is %i" % nerd_age)
```

Result: This user is 34

Run the code above. You'll see that the **nerd_age** variable is a decimal value. Yet, it is printed as a whole number when you print the variable. The reason is that you used *%i* to convert the variable type. A long list of modifiers allows us to convert to another variable type. Below is a condensed table of some common modifiers.

Modifier	Conversion
%d	Decimal
%i	Integer Number
%o	Octal Value
%x	Hexadecimal
%f	Floating Point

Ok, so now I want to print out more than just one variable. How do I accomplish this? Let's use what you have already learned to make this happen. I can practice by outputting the statement:

"Hey Casey is from Nerd Challenges and currently 34 years old".

```
company_name = "Nerd Challenges"
nerd_age = 34.5
print("Hey Casey is from %s and currently %i years
old" % (company_name, nerd_age))
```

Result: Hey Casey is from Nerd Challenges and currently 34 years old

If you got the output above in your run screen, you are now getting the hang of using the legacy string formatter. You can add as many variables as you like and format different or convert various data types. Remember that variables within the print statement must be in the order you specify.

Keeping track of many variables can quickly become cumbersome. There is a much easier way to use the % operator. This approach uses inline variables within the print statement with a **key-value pair**. Building from our last example and morph it into this approach. See below:

```
print("Hey Casey is from %(company_name)s and
currently %(nerd_age)i years old" % {"company_name":
"Nerd Challenges LLC", "nerd_age": 34.5})
```

Result: Hey Casey is from Nerd Challenges LLC and currently 34 years old

Easy enough, this approach eliminates the guessing game when there are multiple variables. You'll probably never write code like this again, but at least you'll be familiar with it if it pops up in older Python 2.7 code.

String Formatting - f-Strings

With Python 3.6 emerged a new method for formatting strings called **string interpolation** or **f-Strings**. This method compounds Python's ability to make code easier to read and write. Since we're just learning Python, this makes our job a lot easier. The syntax is quite simple. You first assign values to your variables. Then you print your statement with the variables inline. The only difference is that you include the letter *f* right before the first quotation mark. Clear your workspace again. Let us revert to one of our earlier examples and morph it using f-Strings:

```
book_day = 1
python_series = 101
```

```
print(f"We are on Day {book_day} of the Python
{python_series} book.")
```

Result: We are on Day 1 of the Python 101 book.

Inside the curly brace brackets, you can evaluate any Python expression. You could even include mathematical operations such as increasing the book day by 1 within the print statement. Try adding + 1 to your previous code:

```
book_day = 1
python_series = 101
print(f"We are on Day {book_day + 1} of the Python
{python_series} book.")
```

Result: We are on Day 2 of the Python 101 book.

Pretty neat, huh! You just incremented the Day by 1 within the print statement. This method is potent. Keep this valuable tool as part of your Python toolbox.

* * *

Introduction to Comments

Most of the code you have written has been only a few lines. As you go into programming further, there is the potential to have programs that contain hundreds of code lines. In large applications, this can approach thousands of lines. There may be an entire team of programmers working on a single project or complicated feature in larger applications. To help with continuity and understanding of the code, programmers use

comments. Comments are situated throughout the program to remind us what is going on. It adds more clarification for others who are reading the code as well.

The syntax for a single-line comment is # (Officially, this is called an Octothorpe, also known as a hashtag). You can comment on code to explain what is going on by inserting a #:

```
# Python does not execute a comment line
print("However, this line prints!")
```

Result: However, this line prints!

The comment tells you what the lines of code below are doing. It doesn't get printed out or even change the output of the code executed. It is helpful for other developers as they try to understand what your code does.

There is only so much room on the first line to provide a comment. Sometimes a comment may span several lines. It may even make more sense to have sentences on subsequent lines. Unlike the single line comment, multi-line comments start and end with three single quotes ''' :

```
''' This is a multi-line comment
The only line that is output by Python is the print
statement
This line is ignored. It is just a way for
developers to document code.
The final program prints The value of x is 3
'''
x = 3
print(f"The value of x is {x}")
```

Result: The value of x is 3

Comments can help with troubleshooting or debugging code. If you have errors, you may consider commenting a portion of your code out. Comments allow you to isolate the problem to make it easier to fix it. Try running the code below:

```
print("Nerd Challenges are awesome!")
print(Nerd Challenges can be challenging)
```

Result: print(Nerd Challenges can be challenging). SyntaxError: invalid syntax

The code doesn't appear to work because of a syntax error. That is weird because I added a line to the code, and now it doesn't work. Let's comment out the portion of the code I just added:

```
print("Nerd Challenges are fantastic!")
# print(Nerd Challenges can be challenging)
```

Result: Nerd Challenges are fantastic!

Since you did not get an error, there is a problem with that particular line of code. Do you see what the problem is?

The issue is that I left out open and close quotation marks in the print statement outside of the string. You can add quotes to the print statement and remove the comment sign. The error I introduced was a simple example. Still, you'll find yourself using comments in more complicated code to help troubleshoot any errors that might emerge. Troubleshooting code in programming is also known as debugging. It is a common term I'm sure you will continue to hear in the future and learn more about towards the end of this book.

* * *

Q&A Review

1.What is an IDE?

1. Integrated Development Environment: Basically a text editor that gives you a bunch of tools to help you write better software.
2. Integer Destructive Environment: A editor that allows you to destroy integers
3. Immutable Development Environment: An editor that is immutable, once you type it's etched in stone.

2.What is the best IDE software?

1. VS Code
2. Pycharm
3. Notepad
4. Sublime Text
5. Notepad++
6. This is a trick question! It's whatever I like the best, for my personal reasons.

3.What is a variable?

1. A variable is code that changes over time by itself, throughout the life of an application.
2. There are no variables in Python, only integers. This is a trick question.

3. A variable allows you to reference data by specifying a property, you can later reference that property in your application. For example my_name = “Casey” would be a variable.

4.How do you print something to the terminal window in Python 3?

1. log(‘I print things like this’)!
2. display(‘Printing cool text’);
3. print(“This doesn’t print anything!“)

5.Why should you use comments?

1. To heckle other developers who look at the code
2. There are no comments in Python
3. To provide meaningful context to a python program

6.Can you have multi-line comments?

1. True
2. False

* * *

Day 1 Challenges

- Kilobyte Challenge: Try and modify the main.py file, instead of saying 'Hi PyCharm' make it say 'Hi, Nerds!'
- Kilobyte Challenge: What happens if you add a comment to the same line as a valid Python command? For example:

```
print("This is a valid python command") # This is a
comment
```

- Megabyte Challenge: What happens if you put a # inside a print statement? For example:

```
print("This shouldn't print # right...?")
```

2

Day 2: Exploring Data Types

Reading Time: 15 minutes

Today, you dive more into data types and operators. I'll take you down the journey of building upon many of the concepts you learned yesterday. The data types you will be exposed to are lists, tuples, sets, and dictionaries. These are also found in many other programming languages. Then I will discuss typecasting, which will allow us to convert the value of one data type to another. Then it's on to learning about the two types of conversions called implicit and explicit type conversion. After today's lesson, you'll understand the primary ways to store data in a Python program.

Lists

Lists are one of the most commonly used data types in Python. They allow us to organize items and keep track of different data types. More than often, a list contains a group of related values. A list is not limited to only one data type either. You can have a list containing a combination of strings, numbers, another list, tuples, Boolean, and even binary.

You can create a list using an open and close square bracket *[ ]*. You put variables of any data type or element within these brackets. You save a list in a variable just like any other data type. Let's go ahead and create our first list that I keep in a variable called **genius_name**. We can assign **geek_names = ["Nikola", "Tesla", "Thomas", "Edison"]**. Notice that all the data types are strings. However, the data types do not have to be the case. I can have numbers included in this list. I am going to add the number 100 just to show this. So **geek_names = ["Nikola", "Tesla", 100, "Thomas", "Edison"]**. Let's print this out just like we print any other variable to see what we get:

```
geek_names = ["Nikola", "Tesla", 100, "Thomas",
"Edison"]
print(geek_names)
```

Result: ['Nikola', 'Tesla', 100, 'Thomas', 'Edison']

There it is, your first list created in Python!

Indexing Lists

The items in a list are indexed, which means that each item is in an ordered position. The first item within any list is index[0], not index[1]. It's important to understand that the first item in any list is in the index[0]! In the list I just created, 'Nikola' is in position 0, 'Tesla' is in position 1, etc. Items in a list are ordered, changeable, and allowed to have duplicates. This concept is referred to as a zero index (0-index). This simply means that whenever you build a list of something in Python, you start with the number zero instead of one. As you learned before, not all programming languages are the same. Some languages, such as R, use 1-indexing, which means when a list is created in R, you start with the number 1.

In the example, I printed out the entire **geek_names** list. Sometimes it is beneficial to print a whole list of items. You, however, may want to select one item within a list. This is where your knowledge of indexing comes in. If you wanted to print out the 4th item in a list, how do you do that? It's actually quite simple. You print out the variable with brackets, including the index you would like to print. Let's see indexing in action:

```
geek_names = ["Nikola", "Tesla", 100, "Thomas",
"Edison"]
print(geek_names[3])
```

Result: Thomas

See that? I selected one item within the list and printed it out. Also, notice that the 4th item is the third (3) index! Go ahead and try to print out other items in the list using their indexes. But what if you want to print the last item or the second to last item? Using the method earlier, you had to count all the items in the list. The last item in any list is *index[-1]*. To print out the last item in our **geek_names** list, you can use another print statement with *index[-1]*.

```
geek_names = ["Nikola","Tesla", 100, "Thomas",
"Edison"]
```

```
print(geek_names[-1])
```

Result: Edison

If I decided to print out the second to last item, I could simply replace the *index[-1]* with *index[-2]*. Sometimes there is also a need to know the number of items within a list. This is where the function *len()* comes in handy. Using this function, you can determine the length or number of items in a list. To find out the length of **geek_names,** I can do the following:

```
geek_names = ["Nikola", "Tesla", 100, "Thomas",
"Edison"]
print(len(geek_names))
```

Result: 5

The number 5 is the total number of items in the **geek_names** list. Now that you know the length, you can select a range of items using *[X:Y]* before the variable. **X** is the first index, and **Y** is the last index. When you select a range of items, the last number or **Y** is not returned. Let's now print out the list except for the last item, "Edison".

```
geek_names = ["Nikola", "Tesla", 100, "Thomas",
"Edison"]
print(geek_names[0:4])
```

Result: ['Nikola', 'Tesla', 100, 'Thomas']

If you wanted to go all the way to the end and print Edison out, you could simply replace **[0:4]** with **[0:5]**. It seems easy enough, except that you may not know the length of the list. Plus, think about what happens if a list were to change. For instance, you could always add or remove items from a list. Hardcoding a number to print out is probably not the best option. A more suitable alternative is using *[X:]* or for this specific case **[0:]**.

```
geek_names = ["Nikola", "Tesla", 100, "Thomas",
"Edison"]
print(geek_names[0:])
```

Result: ['Nikola, 'Tesla', 100, 'Thomas', 'Edison']

Okay, now let us quickly print from "Tesla" to "Thomas" for some additional practice. Here's the code:

```
geek_names = ["Nikola", "Tesla", 100, "Thomas",
"Edison"]
print(geek_names[1:4])
```

Result: ['Tesla, 100,' Thomas']

Now I don't like having a number in this list. It just seems out of place since it's the only numeric value. How could we change that numeric data type to a string similar to the other items? I can do this since we know the index of the number **100**.

```
geek_names = ["Nikola", "Tesla", 100, "Thomas",
"Edison"]
geek_names[2] = "Einstein"
print(geek_names)
```

Result: ['Nikola', 'Tesla', 'Einstein', 'Thomas', 'Edison']

You have replaced the number **100** with **"Einstein"**. Einstein is now the *index[2]* of the **geek_names** list. You can replace any other items within the list by just knowing the index. You can add any data type you want inside a list, as I stated earlier.

You can also look for an item by name to find the index location. I can search through the list to find the first index location for "Tesla".

```
geek_names = ["Nikola", "Tesla", 100, "Thomas",
"Edison"]
print(geek_names.index("Tesla"))
```

Result: 1

The number **1** represents *index[1]*, where the value Tesla is found. If you try searching for a value, not in the list, an error is returned in the console window.

List Functions

In addition to being able to replace items, there is a multitude of functions to modify lists. Instead of going through them all here, I provide a few lists to help set up your environment so you can try a few functions from the table below:

```
nerd_names = ["CJ", "Albert" , "Kelv", "Prodigy",
"Rico", "Jamison"]
geek_names = ["Nikola", "Tesla", "Thomas", "Edison"]
nerd_ages = [8, 12, 25, 13, 15, 18]
geek_ages = [1, 16, 14, 22, 10, 31]
nerd_pop = ["pop1", "pop2", "pop3"]
```

Function	Description	Example	Result
Extend	Allows you to append a list onto another list	nerd_names.extend(geek_names)	Add geek_names list to the end of nerd_names list
Append	Adding an item to the end of the list	nerd_ages.append(70)	Add the number 70 at the end of the nerd_ages list
Insert	Add an item anywhere in the list	geek_names.insert(1, "Newton")	Inserts Newton in Index 1
Remove	Remove an element within the list	nerd_names.remove("Jamison")	Removes Jamison from the nerd_names list
Clear	Clear an entire list	geek_ages.clear()	Clears all items from geek_ages list
Pop	Pop the last item off of a list	nerd_pop.pop()	Results in pop 3 being removed off the nerd_pop list
Count	Count the number of the same elements in the list	nerd_names.count("Albert")	Results in number 1 indicated Albert appears once
Sort	Sort the list in alphabetical order or ascending	geek_names.sort()	Results in the list being alphabetized or ascending order
Reverse	Reverse order of list	nerd_ages.reverse()	This puts the list in reverse order
Copy	Allows the user to copy a list to another variable	nerd_ages2 = nerd_ages.copy()	Copies nerd_ages list to nerd_age2 variable

Tuples

A **tuple** is closely similar to a list in Python. It is a type of data structure that acts as a container to store different values. Multiple data types can be stored within a single tuple, just as a list. When you create a tuple, the ordering of the items in the tuple is preserved.

The biggest thing to remember with tuples is that their contents cannot be later changed or modified. This characteristic of tuples is known as **immutable**. Once we assign a tuple to a variable, there are no changing values within that tuple.

Therefore, if you are working with any data with values that must be changed, it's recommended that you use a list! The contents within a tuple are permanent. Tuples are generally used when data in your program is not expected to change.

A tuple is denoted by open and close parentheses *( )*. You can put variables of any data type or element within these parentheses.

Let's create your first tuple with a collection of shapes. I assign the variable **my_shapes = (Circle, Rectangle, Square, Triangle)**. These shapes that I am working with within the program are permanent. I have no plans to add, remove, or modify any of them from our tuple. They are essentially fixed. Let's print out your first tuple to see what we get.

```
my_shapes = ("Circle", "Rectangle", "Triangle",
"Square")
print(my_shapes)
```

Result: ('Circle', 'Rectangle', 'Triangle', 'Square')

Data within Tuples

Accessing data or items in tuples has the same syntax as lists. You use square brackets to identify the index of the data being selected. Say, for instance, you would like to print out the shape **"Rectangle"**. Your code would simply be print**(my_shapes[1])**, the 2nd value in the **my_shapes** tuple. But hey, what if you decide instead of a **"Rectangle"** we prefer a **"Trapezoid"**? You have easily changed values in a list before. Let us see what happens when you try the following:

```
my_shapes = ("Circle", "Rectangle", "Triangle",
"Square")
my_shapes[1] = "Trapezoid"
print(my_shapes)
```

Result: SyntaxError: invalid character '"' (U+201C)

That looks a bit messy! I don't think we have seen any error messages like this before. I did talk about how tuples are immutable. This means values in a tuple cannot be changed for anything. You tried to change a **"Rectangle"** to a **"Trapezoid"**. This is not allowed when working with tuples. That brings us to another conclusion as well. All those list functions from our table discussed earlier do not work with tuples because they involve changing their contents.

One last topic to review with tuples is datatypes. There might come a time when you would like to store more than just a string or a number value in a tuple. You can even keep a list inside of a tuple and vice versa. Take a look at the code below:

```
my_shapes = ("Circle", "Rectangle", "Triangle",
"Square", [0, 4, 4, 3])
print(my_shapes)
```

Result: ('Circle', 'Rectangle', 'Triangle', 'Square', [0, 4, 4, 3])

I just created a tuple with a list embedded into it!

Sets

A **set** is another datatype worth discussing. It is used to store unordered data and where duplicates are not required. Think of a single dice or a die. A die has six possible values. If rolled, it can only have one value once, so no number of dots on one die is duplicated. I can represent the die as a set. I can enclose the die's values in curly braces *{ }*:

```
my_die = {1, 2, 3, 4, 5, 6, 1, 2}
print(my_die)
```

Result: {1, 2, 3, 4, 5, 6}

In the output notice, our set does not contain any duplicates. The set only stores a value once. You can add the same 6 die values a hundred times, which won't change our output. In this example, 6 values are in numerical order. This is not always the case in sets. You should **never** expect the values are stored in any order. Values within a set should always be treated as arbitrary. To prove this, update the previous example with the following:

```
my_die = {6,7,8,9}
print(my_die)
```

Result: {8, 9, 6, 7}

There is also a way to search through the set for a specific value. If you wanted to know if the number 3 is in our my _die set, you could write:

```
my_die = {1, 2, 3, 4, 5, 6, 1, 2}
print(3 in my_die)
```

Result: True

Within the print function, you search for the number **3** in the set. **TRUE** means the number **3** was found in the set. If we looked for the number **7**, the output would yield **FALSE. TRUE** and **FALSE** are Boolean values returned when an expression is evaluated. In this case, you consider whether or not a number is part of a set of numbers.

Set Functions

Some other useful functions or methods can be used when working with sets. Clear your code and set up your environment with the following sets:

```
my_vowels = {"a", "e", "i", "o", "u"}
my_alpha = {"a", "b", "c", "d", "e"}
```

The first function to look at is called **intersection**. The intersection method is used to determine if there are any items in common among the two sets. You can use the two sets you created to illustrate this:

```
print(my_vowels.intersection(my_alpha))
```

Result: {'a', 'e'}

The second function to look at is called **difference**. The difference method determines if any items are different between the two sets you created. This example shows how the **difference** method can be used among the two sets:

```
print(my_vowels.difference(my_alpha))
```

Result: {'o', 'i', 'u'}

The last function is called **union**. The union method joins two different sets together. You can use the two sets created earlier to show what happens when both are joined together:

```
print(my_vowels.union(my_alpha))
```

Result: {'d', 'c', 'i', 'o', 'e', 'u', 'a', 'b'}

Dictionaries

Dictionaries are similar to a list. They store a collection of ordered objects. They are different than the other data structures in that they are not indexed by numbers but by keys. The syntax is quite similar to a set, whereas you would use curly braces *{ }* to denote a dictionary. Unlike sets, each value within a dictionary has a key-value pair. So each pair have at least a **key** separated by a colon followed by a **value**. A comma separates a pair. Let's create our first dictionary stored in the variable **day_conversion**:

```
day_conversion = {
    "Mon": "Monday",
    "Tues": "Tuesday",
    "Wed": "Wednesday",
    "Thurs": "Thursday",
    "Fri": "Friday",
    "Sat": "Saturday",
    "Sun": "Sunday",
}
```

An important thing to pick up on is the indentation. Each

key-value pair is indented under the variable name. The first column underneath the variable is known as the key. Mon, Tues, Wed, and etc. are all **keys**. The next column holds the values to each of those keys. Essentially, we are saying that a unique key represents each value. Keys must always be different and are not duplicated in a dictionary. The value of those keys, however, can be the same. It is possible to have two keys with the same value. Let's print this dictionary out and also print out the value to the key "Wed":

```
print(day_conversion)
print(day_conversion["Wed"])
```

Result: {'Mon': 'Monday', 'Tues': 'Tuesday', 'Wed': 'Wednesday', 'Thurs': 'Thursday', 'Fri': 'Friday', 'Sat': 'Saturday', 'Sun': 'Sunday'}

Wednesday

I just printed out the entire dictionary with all the key-value pairs. Then I was able to print the value "Wednesday" by using the key "Wed". This can be done for any of the keys in our dictionary. Keys are not limited to strings either. You can use numbers, tuples, and lists too. A similar dictionary might have the months of the year. If you print a key not found in a dictionary, an error message is returned if using the previous method. The *get( )* method allows our program to run without being halted if the key isn't found. Let's rewrite our code with the *get( )* method and type an erroneous key:

```
print(day_conversion.get("not_in_dictionary"))
```

Result: None

The value returned from a key that isn't in the dictionary is now just none. If I replace "not_in_dictionary" with a key in the dictionary, it returns that key's value just as before. This method is better when printing out dictionaries or looping through a list of dictionaries where the key may or may not be found.

Dictionary values cannot be changed in the same manner as lists. There are no indexes, so the only way to change values within its contents is by using a key. Let's try to change the value of **"Sunday"** to **"Funday"** by using its key to see what we get:

```
day_conversion = {
    "Mon": "Monday",
    "Tues": "Tuesday",
    "Wed": "Wednesday",
    "Thurs": "Thursday",
    "Fri": "Friday",
    "Sat": "Saturday",
    "Sun": "Sunday",
 }
day_conversion["Sun"] = "Funday"
print(day_conversion)
```

Result: 'Fri': 'Friday', 'Sat': 'Saturday', 'Sun': 'Funday'

Look at that you now have Funday instead of Sunday. When modifying dictionaries, you must target the key, not the index. Simple enough.

You can use dictionaries to represent objects. If we wanted to store information from our user Casey. You could call the variable or object **caseydict** and capture characteristics regarding his name and job information. That dictionary might look something like this:

```
caseydict = {
    "name": {
        "prefix": "Mr.",
        "first": "Casey",
        "last": "Gerena",
        "suffix": "Jr"
    },
    "job": {
        "company": "Nerd Challenges LLC",
        "position": "Lead Nerd",
        "location": {
            "address": "123 NC Blvd",
            "city": "Tampa",
            "state": "Florida",
            "zip": 33601
        }
    }
}
print(caseydict)
```

Result: {'name': {'prefix': 'Mr.', 'first': 'Casey', 'last': 'Gerena', 'suffix': 'Jr'}, 'job': {'company': 'Nerd Challenges LLC', 'position': 'Lead Nerd', 'location': {'address': '123 NC Blvd', 'city': 'Tampa', 'state': 'Florida', 'zip': 33601}}}

Try running this code for yourself. It becomes clear how storing key-value pairs in a dictionary can apply to real-world scenarios like constructing databases or storing important user information. This format is identical to JavaScript Object Notation (JSON).

Type Casting

Another helpful tool in Python to learn is **type conversion**. Type conversion involves converting one data type to another data type. Variables can store values of all different data types. Certain operations may requ ire a variable's value to be a specific data type. This is where type casting or type conversion is necessary. There are two types of type conversion:

- Implicit Type Conversion
- Explicit Type Conversion

Implicit Type Conversion

Implicit type conversion automatically converts data types from one type to another. Python performs implicit type conversion, and it does not require any interaction from a user to do the implicit type conversion. When you assign a value to a variable implicit type conversion occurs. When you set the variable **nerd_name = "Casey"**, Python automatically converted this value into a string data type.

You can check your variables' data type using the *type( )* method. The *type()* method gives you an idea of how variables are implicitly typed cast. Let's check out some of the data types I discussed earlier by using this method:

```
nerd_name = "Casey"
nerd_age = 34
```

```
pi = 3.1415926535897

nerd_names = ["CJ", "Albert", "Kelv", "Prodigy",
"Rico", "Jamison"]
my_shapes = ("Circle", "Rectangle", "Triangle",
"Square")
my_die = {1, 2, 3, 4, 5, 6, 1, 2}

print(type(nerd_name))
print(type(nerd_age))
print(type(pi))
print("\n")
print(type(nerd_names))
print(type(my_shapes))
print(type(my_die))
```

Result: <class ‘str’>
<class ‘int’>
<class ‘float’>

<class ‘list’>
<class ‘tuple’>
<class ‘set’>

I forgot to include the dictionary datatype that we just discussed! Let us print out the data type for a dictionary assigned to the variable **day_conversion**.

```
day_conversion = {
    "Mon": "Monday",
    "Tues": "Tuesday",
    "Wed": "Wednesday",
    "Thurs": "Thursday",
    "Fri": "Friday",
```

```
    "Sat": "Saturday",
    "Sun": "Sunday",
}
print(type(day_conversion))
```

Result: <class 'dict'>

Python can also do math calculations on numbers of two different data types. See what happens when I add **nerd_age = 34,** which is an integer, and **pi = 3.141592653589793,** which is a float together:

```
nerd_age = 34
pi = 3.1415926535897
sum = nerd_age + pi
print(sum)
print(type(sum))
```

Result: 37.1415926535897 <class 'float'>

You'll notice that the final value or sum is the data type **float**. Python converted this for you automatically. It wouldn't make any sense for the sum to be an integer. Doing this would cut off the decimal portion, yielding an incorrect value. Python helps us out, so this does not occur by implicit type conversion.

Explicit Type Conversion

Explicit type conversion uses predefined functions to convert a data type from one to another. Below is a list of standard predefined functions; however, there are many others.

- **str( )** - converts a value to a string

- **int( )** - converts a value to an integer number
- **float( )** - converts a value to a float/decimal number
- **list( )** - converts a value to a list
- **tuple( )** - converts a value to a tuple
- **set( )** - converts a value into a set

When you concatenated strings on Day 2, you received an error if you tried printing a string and integer in the same print statement. With explicit type conversion, you can simply convert an integer or number to a string so that it prints:

```
nerd_name = "Casey"
nerd_age = 34
print(nerd_name + " is of the age " + str(nerd_age))
```

Result: Casey is of the age 34

I used explicit type conversion to print out a statement that included an integer. An error occurred if I did not use the predefined function *str( )* because you cannot concatenate a string with a number. This is one scenario where explicit type casting is practical. As you practice Python coding, you will encounter other situations.

* * *

Q&A Review

1.How do you convert a number to a string, explicitly?

1. By using the *str()* method
2. By using the *int()* method
3. By using the *.toString()* method
4. By using the *.toInteger()* method

2.Tuples are immutable?

1. True
2. False

* * *

Day 2 Challenges

- Kilobyte Challenge: Create a list of fruits with ‘apple‘ and ‘banana‘. Once that list is created, add a new fruit to the list using the append method.

3

Day 3: Operators

Reading Time: 3 minutes

Today's lesson is all about Python operators. I go into more depth and build upon some of the basic arithmetic operators I talked about on Day 2 to perform math operations. Python operators can be consolidated into the following groups: Arithmetic, Assignment, Bitwise, Comparison, Identity, Logical, and Membership Operators. The list below is a general description of each:

- **Arithmetic** - used to perform math operations
- **Assignment** - used to assign values to variables
- **Bitwise** - used to compare numbers in binary
- **Comparison** - used to compare two variables or values
- **Identity** - used to determine if a value is or is not something (i.e., **True** or **False**)
- **Logical** - used with conditional statements
- **Membership** - used to identify if a value is within a set or sequence

Arithmetic Operators

Let us refresh our knowledge of what you learned yesterday by using the table below. There is one additional operator I added to the table called floor division. Floor division takes a quotient and rounds down to the lowest integer. For simplicity let us set the variable x = 1 and y = 3. Check out the results in the last column of the table to see what results you receive from using each operator.

Operator	Name	Example	Results
+	Addition	x + y	4
-	Subtraction	x - y	-2
*	Multiplication	x * y	3
**	Exponent	x ** y	1
/	Division	x / y	0.333333
%	Modulus	x % y	1
//	Floor Division	x//y	0

Assignment Operators

You have assigned values to variables using the "=" **assignment operator**. A few more assignment operators are used throughout the rest of the book. Don't focus too heavily on these operators now. You'll use these operators when you learn for and while loops in subsequent days.

Operator	Name	Example	Solution
=	Assignment	x = 5	x = 5
+=	Add & Assignment	x += 3	x = x + 3
-=	Subtract & Assignment	x -= 2	x = x - 2
*=	Multiply & Assignment	x *= 4	x = x * 4
/=	Division & Assignment	x /= 6	x = x / 6

Bitwise Operators

Bitwise operators compare two binary numbers with one another. These would only be used when using binary numbers, including 0's and 1's. There is no need to go in-depth into these now. All the numbers in this book are whole or decimal.

Comparison Operators

Comparison operators are used in most of the programs you create in Python, especially if they contain if-else statements, while, and for loops. You can compare two variables with comparison operators and then execute a set of instructions if they meet specific criteria or conditions. You will learn more about comparison operators on Day 5; for now, just know these are the symbols and how to use them.

Operator	Name	Example
==	Equal	x == y
>	Greater Than	x > y
>=	Greater Than or Equal to	x >= y
<	Less Than	x < y
<=	Less Than or Equal to	x <= y
!=	Not Equal	x != y

Identity Operators

Similar to a comparison operator. An **identity operator** examines two variables to see if they are the same. The easiest way to show this is with a small script comparing three different variables. The two identity operators are **is** and **is not**:

```
a = 7
b = 3
c = 7

print(a is b)
print(a is c)
print(a is not b)
```

Result: False, True, True

The result of the first print statement is **FALSE** because **a = 7** is not equal to **b=3**. The second print statement is **TRUE** because **a = 7** equals **c = 7**. The third print statement is **TRUE** because **a =7** is not equal to **b =3**. The response from an identity operation is always either **TRUE** or **FALSE**.

Logical Operators

Logical operators are used when dealing with more than one conditional statement. It allows us to include or look at two conditions at one time when evaluating a statement. The three logical operators below are used more on Day 6.

- **and** - returns a response if both conditions are met
- **or** - returns a response if one or more conditions is met
- **not** - returns the opposite of the response (i.e., if statement is **TRUE**, return **FALSE** and vice versa)

Membership Operators

The last type of operator is **membership**. Think of it just like an employee in a company. An employee who works for a company is regarded as a member. The same thing goes with items or variables in Python. You receive a **TRUE** value for a specified value found within a group. You receive a **FALSE** if that specified value is not in that group. Using membership consists of the operators **in** and **not in**. Let's figure out who is an employee of the Nerd Challenges company:

```
nerd_company = ["Casey", "Ricardo", "Marilyn", "Mia"]
print("Ricardo" in nerd_company)
print("Jessica" in nerd_company)
print("Ricardo" not in nerd_company)
print("Jessica" not in nerd_company)
```

Result: True, False, False, True

So, from the code above, I saved a list of all the employees who work for Nerd Challenges. To see if Ricardo or Jessica are members, I used a print statement with a membership operator to check. The results of the operators are shown in the output above. Ricardo is, in fact, and member of the company, but Jessica is clearly not!

* * *

Q&A Review

1.What operator is used to multiply two numbers together?

1. +
2. -
3. &
4. *

2.There are comparison operators in Python, which allow you to compare integers.

1. True
2. False

* * *

Day 3 Challenges

- Kilobyte Challenge: Create two variables with an integer stored in each. Then try to multiply the variables together.

4

Day 4: User Interaction

Reading Time: 3 minutes

Printing out statements is all fine and dandy but doesn't meet all the criteria for an actual application. What if I don't just want our program to print out statements? Many programs that you interact with daily require additional inputs from their users. Today you learn to incorporate user interactions into the program by prompting the user for information.

Remember when you learned how to print the statement **"Welcome to the Nerd Challenges Python101 Course"**. All you did was print the variable **nerd_welcome,** which contained the string **"Welcome to the Nerd Challenges Python101 Course"**. This works, but what if you wanted to welcome a person to the course? You would then need to ask that person for their name.

To request input from a user in Python, you use the *input( )* function. This function will read in a line of input from the user. You can then save that user input into a variable for later use. To prompt a user for their name and print out that name, use the following code:

```
name = input("Enter your name: ")
print("Welcome to Nerd Challenges " + name)
```

Result: *Enter your name: John*

Welcome to Nerd Challenges John

After running the code above, you prompted the user for their name. The user then entered the name **"John"** into the program. The input function then saved that name into a variable called **name**. In the print statement, you printed out the **name** variable. The same variable was just initiated by the user.

Excellent, you now know how to ask the user for their input. Continue with making your program more interactive. You can practice by asking the user for a few different things. Ask the user for the following information:

1. Name
2. Title of the book being read
3. Reason for learning Python

Be creative because you'll be printing these values out as you did a minute ago. Once you are done, see the code below for a sample of what your code may look like. There are no right or wrong answers when it comes to coding. There are many ways to accomplish the same thing.

```
print("Welcome to the Nerd Challenges!")
name = input("Enter your name: ")
print("Hey " + name + ", great to have you aboard.")
title_of_book = input("What is the title of the book
you are reading? ")
```

```
print(title_of_book + " is a great resource to learn
Python.")
python_reason = input("What is your reason for
wanting to learn Python? ")
print(python_reason + " is an excellent reason to
learn how to program in Python")
```

Result:

Welcome to the Nerd Challenges!

Enter your name: Casey

Hey Casey, great to have you aboard.

What is the title of the book you are reading? Python 101

Python 101 is a great resource to learn Python.

What is your reason for wanting to learn Python? To improve my skills

To improve my skills is an excellent reason to learn how to program in Python

You can also prompt the user for numbers using the same *input( )* function. If a number is entered in Python, it is automatically saved as a string. You can easily convert that string into a number using explicit type conversion. Remember what you learned on Day 3? Review the example below before you proceed to dive deeper into the realm of Python.

```
course_number = input("Input the course number: ")
print("The course number + 1 is")
print(int(course_number) + 1)
```

Result:

Input the course number: 101

The course number + 1 is:

102

Today was a short lesson; however it contains some fantastic concepts, be sure to practice them! Wait to see what we have in store for tomorrow's lesson.

* * *

Q&A Review

1.What Python function is used to prompt the user for input?

1. print()
2. input()
3. read()
4. write()

2.When a user responds to an input function with a number, the variable is saved as a string.

1. True
2. False

* * *

Day 4 Challenges

- Kilobyte Challenge: Write a short story that uses inputs from the user, by asking the user questions.

5

Day 5: If-Else Statements, While, and For Loops

Reading Time: 10 minutes

Today is dedicated to using Python if-else statements, while, and for loops. I will review all of these concepts and how to use them in your program. These concepts apply to most other programming languages as well. There are subtle differences that stem from the syntax in which they are used. Nonetheless, you'll be able to take these concepts to implement in other languages.

If-Else Statements

In the real world, you make decisions all the time. Each decision you make has some type of response or an expected result. You can write programs using if-else statements to execute actions based on user inputs. If-else statements are ideal to use in Python for decision-making.

An **if-else statement** starts with *"if"* and has a statement that is evaluated. If that statement is **TRUE**, it will continue to execute the subsequent operations. If that statement is **FALSE**, the program will exit out of the if-else statement, not executing the following operations. You can use the comparison operators that you learned about on Day 3.

Operator	Name	Example
==	Equal	x == y
>	Greater Than	x > y
>=	Greater Than or Equal to	x >= y
<	Less Than	x < y
<=	Less Than or Equal to	x <= y
!=	Not Equal	x != y

I want you to write an if-else statement that takes a user's input to see if they are old enough to go on a ride at a theme park. Theme parks generally have age limits for their rides. In this statement, you need to assign values to variables for the rider's age and the age limit. Those numbers can be compared, and if the condition is met (i.e., the rider is over the age limit), a statement is printed out that the rider is old enough to go on the ride. The if-else statement would look something like this:

```
rider_age = 18
age_limit = 16
if rider_age >= age_limit:
    print("Congrats! You are old enough to go on
    this ride.")
```

Result: Congrats! You are old enough to go on this ride.

Since the rider's age is 18 and older than the age limit, the condition of the if-else statement has been met. The print statement can be executed to print that the rider is old enough to get on the ride. The >= operator essentially refers to the **rider_age** variable being greater or equal to the **age_limit**. If you change the **rider_age** variable to **13**, the required conditions are false or not met. This results in the output not executing the print statement:

```
rider_age = 13
age_limit = 16
if rider_age >= age_limit:
    print("Congrats! You are old enough to go on
    this ride.")
```

Result: Process finished with exit code 0

Notice the condition was not met, so nothing was returned. Wait a minute, how does the user of the program know that the condition has not been met without any visual indication? Someone would let you know if you were not old enough to go on a ride in real life. You are now going to add a print statement to let the rider know they are not old enough if they don't meet the required conditions:

```
rider_age = 13
age_limit = 16
if rider_age >= age_limit:
    print("Congrats! You are old enough to go on
    this ride.")
else:
    print("Sorry, you are not old enough to go on
    this ride!")
```

Result: Sorry, you are not old enough to go on this ride!

Since you added the *else* to the if-else statement, the program returned a result when the required conditions were not met. It is good practice to let a user know when the expected input does not meet expected conditions or requirements.

Say you now want a rider under the age limit to know that he is very close to reaching the age limit. Maybe that rider is only one year from being able to go on a ride. You could put another condition into the if-else statement using *elif*. Now when the rider's age is over 13, the program will notify them that they are closer to the age required to go on the ride:

```
rider_age = 15
age_limit = 16
if rider_age >= age_limit:
    print("Congrats! You are old enough to go on
    this ride.")
elif rider_age >= 13:
    print("You are almost old enough to go on this
    ride.")
else:
    print("Sorry, you are way too young to go on
    this ride!")
```

Result: You are almost old enough to go on this ride.

That was easy. The program still has the last condition to capture any inputs not meeting any of the two above conditions. Your program should always have at least an *if* and *else* for if-else statements. The *elif* is optional but provides the ability to check multiple conditions. The if-else statement always starts checking the first or top condition then proceeds to the next until it ends.

There are cases where a condition has two requirements.

Using the previous example, if age is not the only thing required to get on a ride, it is necessary to consider other factors, like height. Someone too tall could very well not be able to fit into the ride vehicle. Thankfully there are logical operators such as *and, or, not* to help with this. The first condition we have should check the age of the rider as well as the height.

```
rider_age = 20
age_limit = 16
tall = True

if rider_age >= age_limit and tall:
    print("Congrats! You are old enough and tall
    enough for this ride.")
elif rider_age >= 13 and tall:
    print("You are almost old enough to go on this
    ride, but you are tall enough!")
else:
    print("Sorry, you are way too young or way too
    short to go on this ride!")
```

Result: Congrats! You are old enough and tall enough for this ride.

The logical operator *and* is used to check for two factors: a rider's age and a rider's height. Logical operators allow us to ensure two requirements of a condition are met or even if just a single condition is met. The variable tall was set to **TRUE** but could easily be changed to **FALSE**. Examine the output when the **tall** variable is now set to **FALSE**:

```
rider_age = 20
age_limit = 16
tall = False
```

```
if rider_age >= age_limit and tall:
    print("Congrats! You are old enough and tall
    enough for this ride.")
elif rider_age >= 13 and tall:
    print("You are almost old enough to go on this
    ride, but you are tall enough!")
else:
    print("Sorry, you are way too young or way too
    tall to go on this ride!")
```

Result: *Sorry, you are way too young or way too tall to go on this ride!*

There is one other logical operator I've been neglecting to mention called *or*. The operator *or* looks to see if at least one of the requirements has met the condition. For instance, let's say you would like to let someone go on the ride only if they are over 16 or tall. You could write your code as follows using the *or* and <= operators:

```
rider_age = 18
age_limit = 16
tall = True
if rider_age <= age_limit or tall:
    print("Congrats! You are either older than 16 or
    tall enough to ride.")
else:
    print("Sorry, you are either younger than 16 or
    not tall enough to ride!")
```

Result: *Congrats! You are either older than 16 or tall enough to ride.*

Notice that only one of the requirements was met for the condition. The rider is over 18, so it would appear he cannot go on the ride at first glance. The *or* operator, however, allows the statement to check if the variable is **TRUE**. The variable **tall** is **TRUE**, so this condition is met. The program proceeds to print out the statement, **"Congrats! You are either under 16 or tall enough to ride."** Only one of the arguments needs to be **TRUE** to be valid for ***or*** the operator.

While Loops

While loops check to see if a condition is met to execute an operation. A while loop continues to execute operations as long as the required condition is met. Understanding that a while loop will not end while the required condition is met is critical. So, if you decide to create code where the while loop condition is always valid, that while loop will continue to run, and your program will never stop. For that to not happen, you must ensure you insert code to end the loop eventually. Assign the variable **x** = **3**. While the variable x is less than the number **10,** print out the value of the variable **x**. Check out the code below:

```
x = 3
while (x < 10):
    print(x)
    x = 15
```

Result: 3

Since the value **3** is less than the condition **x<10,** the value of **x** = **3** is printed out. If I left the code **x** = **15** off, the program would continue to run over and over, printing out **x** = **3**. It would essentially turn into an infinite loop with no end. You do not ever want this to happen.

The other thing you would never want is your program to assign an arbitrary value to one of the variables in the loop. Not knowing what value is being passed into your while loop is dangerous. It could also cause an endless loop.

Start to get into a "what if" mindset when programming. Always think about what happens if a specific value or condition is present. Your program needs to be able to handle different situations. As programmers, you have to think about these. While loops are the first time you had to think "what if". I assure you there will be more as we continue on.

Let us continue to build upon your while loop. You now know you want to print out every x less than **10**. At the same time, you decide to increment by the number 1 till you get to 10. That might be cool because now you are printing out numbers **4** to **10** using only a few lines of code. Try using the code below to do this:

```
x = 3
while (x < 10):
    x = x + 1
    print(x)
```

Result:

4
5
6
7
8
9
10

In this code, you started with the number **3,** which is less than **10**, then executed **x = x +1**. This sets **x = 3 + 1**, which returns the value **4**. The value **4** is now saved in variable **x**. This is also the value that is printed out on the next line. Wait a minute; this process starts again because **x** is now equal to **4**, which is still less than **10**. The code under our while loop now prints out **5**. The process continues until you eventually get to the number **10**! So, you were able to print numbers **4** to **10** and exit out of the loop once the value became larger than **10**. When **x** was equal to **10**, the loop ended. Incrementing a number by one is a great way to end a while loop.

You can clean this code up by using one of the assignment operators discussed earlier. Instead of using *x = x + 1* replace that with *x +=1*. From now on, you can use this notation to simplify your code:

```
x = 3
while (x < 10):
    x += 1
    print(x)
```

Result:

4
5
6
7
8
9
10

You still need to let the user know when the loop has ended. In the if-else statement, you let the user know when no condition was met. The same thing should go for loops. Print out a message outside the while loop to indicate that the loop has finished.

```
x = 3
while (x < 10):
    x += 1
    print(x)
print("This loop has now been terminated!")
```

Result:

4
5
6
7
8
9
10
This loop has now been terminated!

For Loops

For loops work similarly to while loops. In my opinion, they can be easier to use. For loops can loop through various data types, including strings and lists. These loops allow you to loop through all kinds of data types.

I always wanted an easy way to print individual letters of a string in a program but never knew how. My first thought was always to use a while loop. While this is possible, it isn't the best way. For loops are quicker at handling tasks such as this without the need for multiple lines of code. So how do I print out the individual letters for **"Nerd Challenges LLC"**? Well, I just need to loop through a variable that has the string stored:

```
company = "Nerd Challenges LLC"
for letter in company:
    print(letter)
```

Result:

N

e

r

d

C

h

a

l

l

e

n

g

e

s

L

L

C

A letter is printed out for every letter in the variable, **company**. This can be done with any string. Let's try this without using any string variables.

```
for letter in "Nerd Challenges LLC":
    print(letter)
```

Result:

N

e

r

d

C

h

a

l

l

e

n

g

e

s

L

L

C

I get the same output as before without assigning a value to a variable. How about a list of names? If I use a for loop for a list of names, the loop looks at each of the items in the list:

```
geek_names = ["Nikola", "Tesla", "Thomas", "Edison"]
for name in geek_names:
    print(name)
```

Result:

Nikola

Tesla

Thomas

Edison

That was cool. Each name from the list of names was printed out. The same code can be used to print out tuples or sets:

```
geek_names_list = ["Nikola", "Tesla", "Thomas",
"Edison"]
geek_names_tuple = ("Nikola", "Tesla", "Thomas",
"Edison")
geek_names_set = {"Nikola", "Tesla", "Thomas",
"Edison"}
```

For loops, work with numbers as well. Remember our while loop where you printed out numbers **4** to **10**. That required almost 5 lines of code. What if you could only use two lines of code to accomplish the same thing? Well, you can, and here's how:

```
for number in range(4,11):
    print(number)
```

Result:

4

5

6

7

8

9

10

The **range** function gives us the ability to select a range of numbers. The first number is included, and the last number is excluded. **11** is used because we wanted to include all numbers from **4** to **10**. If you only specify one value, such as **range(11)**, the range would be from **0** to **10**.

* * *

Q&A Review

1.What type of statement in Python can be used to make a decision?

1. operators
2. if-else statements
3. print statements
4. data types

2.Suppose you have the below code block and it is executed, what will you see on the terminal?

```
rider_age = 13
age_limit = 16
if rider_age >= age_limit:
  print("Congrats! You are old enough to go on this
  ride!")
```

1. Congrats! You are old enough to go on this ride!
2. Nothing will be outputted
3. Sorry! You are not old enough to go on this ride
4. Invalid, the variables supplied are not valid.

3.In order to use a ‘if‘ statement, you must also use an else.

1. True
2. False

* * *

Day 5 Challenges

- Kilobyte Challenge: Ask a user if they would like to double or triple a number. Then ask for their number. Then double or triple it based on their previous selection.

6

Day 6: Try-Exceptions

Reading Time: 6 minutes

A part of programming is learning how to handle exceptions and errors. An error within the code or even input received by a user can render a program useless. Useless means that the entire program crashes, and you are left looking at an error message. Not all errors should be considered critical enough to stop the whole program from working.

Try and Except Blocks

Try, except, and finally, blocks can help you manage these errors. Errors enable us as programmers to figure out where issues are in code. Users generally do not care to see this additional information about program errors. A simple one-line with an error might be sufficient enough and does not distract a user from proceeding with running the program.

Within the try block, you can test a code block to see if it has any errors. If it does not, the code executes as expected. However, if there is an error, an except block handles the error.

This block is sometimes referred to as an exception. Lastly, there is a finally block that will execute code independent of whether there is an error or not within the try-except blocks.

A standard calculator gives us the ability to conduct almost every type of math operation you can think of. I say almost because if you try to divide any number by zero on a calculator, you will receive an error such as "Cannot divide by zero". If you don't believe me, go ahead and try for yourself.

The same goes for math operations in Python code. A Python program, too, would not be able to execute code that tries to divide a number by zero.

You can always just ensure the code you write doesn't have a number that is being divided by zero. This seems fine, but often a program is interactive and will ask for input from a user. Suppose a user inputs a number for division by zero. In that case, the program will immediately stop with an error message. Try for yourself to divide a number by zero to see the results:

```
result = 5/0
print(result)
```

Result: Traceback (most recent call last) result = 5/0. ZeroDivision-Error: division by zero

The number **5** clearly cannot be divided by the number **0**. The error message below stopped the program while trying to execute that line of code. The error message also indicates the type of error which occurred. Trying to divide by zero yielded a *ZeroDivisionError*, so no code after the program stopped was able to execute. You just broke the program. Thankfully you are learning to use try-except blocks so this won't happen again!

Adding a *try-except* allows you to handle an error such as dividing by zero. This time add a try block to try dividing 5 by zero, then printing. For except block, you can write an exception that prints out the reason for the error:

```
try:
    result = 5/0
    print(result)
except Exception:
    print("Cannot divide by zero")
```

Result: Cannot divide by zero

The output of the program looks better than what was received before. The program still has an error, but it was still able to execute all of its code without any issues. The output provided the user with the reason why there was an error within the code. The error message printed out is very generic. Also, important to note is that any kind of error in the code will give us the result. What if you just tried printing out a variable that has not yet been assigned a value in the same try-except blocks?

```
try:
   print(unknown_variable)
except Exception:
    print("Cannot divide by zero")
```

***Result:** Cannot divide by zero*

That output is not even close to what you should be expecting. It has the wrong error message. The code is not dividing any numbers by zero. It should be saying something to the effect of a lousy variable or variable not found. I just did this to show you that the except block will consider any errors found in the try block as a general error if the except is an **Exception**. This is why it is necessary to be specific with the error types rather than using the generic Exception.

There are a lot of possible types of errors. Navigate to the website: https://docs.python.org/3/library/exceptions.html#BaseException for a list of all the errors which you could encounter. *NameError* is the exception needed for a missing or unassigned value to a variable:

```
try:
  print(unknown_variable)
except NameError as error:
   print(error)
```

Result: *name 'unknown_variable' is not defined*

Now the exception is **NameError,** and the error for that error type is being printed out. This is how you should be writing your try-except blocks. Some scenarios could require multiple exceptions:

```
try:
    result = 5 / 0
    print(result)
except NameError as error:
    print(error)
except ZeroDivisionError as error:
    print(error)
```

Result: division by zero

In this case, the exception is a **ZeroDivisionError**. However, the program did first check if there were any **NameError** exceptions first. If the result variable was not assigned a variable, it would have resulted in a **NameError** exception. There is no need to print out a custom error message when you know the error type to look for. The *as error* code helps you better illustrate the error. The generic exception can still print out error messages even if you do not know the specific error. Back to our initial example using the *as error* to print out a message for the first error caught by the program:

```
try:
    result = 5/0
    print(result)
except Exception as error:
    print(error)
```

Result: division by zero

The code above was able to catch the exception and print out the standardized message associated with the error type. Try to be specific with your exceptions. If not, there is always the generic Exception class. You haven't learned about classes yet, but I will cover this topic later in the book.

Else Clauses and Finally

The **else clause** can be added to the previous code to execute more instructions after an exception is not found. Using the ZeroDivisionError example, you should print the result only if the exception isn't caught. For that to happen, divide the result by 1 instead of 0 and add an else clause to print the result:

```
try:
    result = 5/1
except Exception as error:
    print(error)
else:
    print(result)
```

Result: 5.0

No exception was caught, so the program printed out the result from the else clause. If you wanted code to execute regardless of whether or not an exception was caught, you could add *finally* to the end of the code. *Finally*, will always execute without taking into consideration exceptions or errors:

```
try:
    result = 5/0
except Exception as error:
    print(error)
else:
    print(result)
finally:
    print("This program runs even if exceptions
    occur.")
```

Result: division by zero. This program runs even if exceptions occur.

Raise

Exceptions are caught when a block of code within try-except is executed. The Python interpreter automatically catches this upon execution. There is a way for us to **raise** an exception manually. You can include a raise exception in the try block to manually trigger the exception. We will take our 5/1 example and manually raise the exception. This example normally wouldn't throw an exception because 5 can easily be divided by 1. I said normally, but this time, I include a raised exception:

```
try:
    result = 5/1
    raise Exception
except Exception:
    print("This is a raised exception!")
```

Result: This is a raised exception!

Notice that the code in the try block doesn't have any actual errors or exceptions. The only code in the try block that would throw an exception is the raise Exception inserted below the result variable assignment. That raise exception triggers the except block.

* * *

Q&A Review

1.How do you handle errors or exceptions within a Python program?

1. if-else statements
2. input function
3. operators
4. try-except blocks

2.What is the term used to describe the joining or combining of strings?

1. Union
2. Intersection
3. Concatenation
4. Join

3.Finally blocks always run

1. True
2. False

* * *

Day 6 Challenges

- Kilobyte Challenge: Ask the user to input a number between 1 - 25. If they input a number above 25, raise an exception. If they input a number below 1, raise a different exception.

7

Day 7: Nerd Challenge 1

Reading Time: 3 minutes

You have already made it through half of the book. In Days 1 to 7, you built the foundation to become a Python programmer:

- **Day 1** – you installed Python and were introduced to variables
- **Day 2** – you learned about data types and intro to basic operators
- **Day 3** – you dove into more advanced operators
- **Day 4** – you learned how to prompt users for input
- **Day 5** – you know how to create if-else statements and loops
- **Day 6** – you built a program that handled exceptions and errors

It is time to put everything together with a small Python project or what I call a Nerd Challenge with all that knowledge. The scenario is simple. I want you to create a calculator able to do simple math operations such as add, subtract, multiply, and divide.

I would also like you to give the user the option to select between a simple and a more advanced calculator, which is still under development. I've provided some guidance below. Remember to break the program into small pieces. Programming does require a lot of trial and error so take your time.

It's go time. Are you going to proceed on with being a Python programmer? If so, then say to yourself, **Challenge Accepted!**

Part 1: Implement Overall Requirements

In this first part, you'll be collecting input from a user. Below are the requirements for fulfilling this part of the challenge

1. Ask the user what their name is
2. Welcome the user to using a program called Python Calculator Program
3. Prompt the user to enter which calculator to use: Simple Calculator or Advanced Calculator
4. If Simple Calculator is chosen: The user should then be asked if they want to add, subtract, multiply, or divide two numbers. Assign the input value to the variable **op**, short for operator. Prompt user for the first number and save the value as variable **num1** for the second variable save value as variable **num2**
5. If the Advanced Calculator is chosen, the program should print out a message stating, **"Advanced Calculator is still under development, please try the Simple Calculator"**
6. If the user inputs a string other than Simple or Advanced, tell the user their input is "Invalid and to try again."

Part 2: Set up the calculator

In the second part, you'll be implementing the logic for the **op** variable

1. Add a nested if-else statement under the Simple calculator to handle the op variable, which is determined by the math operator selected by the user:
2. If the user selects **add**, the program should print out the addition of variables **num1** and **num2**.
3. If the user selects **subtract**, the program should print out the difference of variables **num1** and **num2**.
4. If the user selects **multiply**, the program should print out the product of variables **num1** and **num2**.
5. If the user selects **divide**, the program should print out the quotient of variables **num1** and **num2**.
6. The calculator should handle invalid input

Solution

No peeking ⊠! Check this answer once you've spent at least 2 hours on the challenge.

* * *

```
user = input("What is your name? ")
print("Welcome " + user + " to the Python Calculator
Program!")
calculator = input("Please choose which calculator
to use: Simple or Advanced. ")
```

```
if calculator.lower() == "simple":
    op = input("Would you like to (add, subtract,
    multiply, or divide) two numbers?  ")
    num1 = input("Provide your first number: ")
    num2 = input("Provide your second number: ")
    if op.lower() == "add":
        print(float(num1) + float(num2))
    elif op.lower() == "subtract":
        print(float(num1) - float(num2))
    elif op.lower() == "multiply":
        print(float(num1) * float(num2))
    elif op.lower() == "divide":
        try:
            print(float(num1) / float(num2))
        except ZeroDivisionError as error:
            print(error)
    else:
        print("Invalid op or #, expected a value of
        add, subtract, multiply, divide, number")
elif calculator == "Advanced" or "advanced":
    print("The Advanced Calculator is still under
    development. Please use Simple calculator. ")
else:
    print("The input you chose is invalid. Please
    run program again")
```

8

Day 8: Functions

Reading Time: 9 minutes

A function executes a group of instructions to do one specific thing. Once the function is defined, it can be used anywhere in the program to complete a task.

If you want to do a task that requires a block of code and plan on using it repeatedly, it's probably best to use a function. You would only need to define the function once with all of its code. Then you would be able to use that function throughout your program by just typing the function name.

If your function is composed of multiple lines of code, this can save you a lot of time. Could you imagine having to type in a hundred lines of code more than once to perform the same thing? Well, functions eliminate this need. You only need to type the code once when defining the function, then call the function to reuse the code later.

There are 2 types of functions:

- Built-in Functions
- User-defined Functions

Built-in Functions

Little did you know that you have been using built-in functions throughout the entire course. Built-in functions come with Python and natively support the ability to perform specific tasks. Look at some of the built-in functions you have used in previous lessons:

- print()
- format()
- int()
- float()
- len()
- range()
- tuple()
- set()

The only thing you have to do to use these functions is write the function name and sometimes pass values into it. Take, for example, the common function used to create print statements. You can print by writing the code:

```
print("This is a built-in function.")
```

Result: This is a built-in function.

The print function is already defined internally within Python. Python takes in any value and converts it into a string before printing it out. You can print any value you want, but if and only if you provide the function with a value. This value put into the *print* function is called an argument or parameter.

A function can require one or more parameters to perform the specified task. Arguments are separated by commas. *print( )* does, in fact, accept more than one argument:

```
print("Hello programmer.", "Welcome to Python 101!",
"3rd argument of function", sep = "\n")
```

Result: Hello programmer.

Welcome to Python 101.

3rd argument of function

This print function has a total of 4 parameters. The first three parameters are strings, and the last one is a separator between each string. Since *** was chosen as the separator, it creates a new line between each string or argument! Separators can include any text or escape character. By default, the separator is a space. If a separator isn't specified, space is included between each string argument provided.

User-defined Functions

User-defined functions give you the ability to create your own functions in Python. You can tailor your code for specific use cases. Functions can too be used to break your code into smaller pieces when developing an application. Say you'd like to add features to your program. One feature could be created and separated by a user-defined function. Essentially you can make it easier to develop a program by breaking it into more manageable pieces. This keeps your program running as expected while simultaneously integrating new functionality.

Creating Functions

To create your own function, start off with the word *def*. The syntax *def* is short for the definition of the function. Following that is the function name. All function names are unique, meaning they cannot have the same name as any other function used in the same program. Be as descriptive as possible when considering a function name. This is the name you will call in your program to complete a specific task:

```
def function_name():
```

Parameters or arguments allow the passing of values. Parameters are located between the parentheses. Right after the parentheses is a colon ":" to indicate the start of the function code. Everything else in the function is known as the body. Each function must include a body, or a syntax error will occur. The code above has no instructions for the function, so an error would occur if you tried to run this code in its current state. Let's add a print statement, so this function actually works:

```
def function_name():
    print("My first User-defined Function!")
```

Result: Nothing

Well, that isn't quite what you expected. The program ran but did nothing but show a black screen. That is because you just defined the function. To use a function in a program, you now need to call that function. Calling refers to using the function name in your program. Let's try again with a call:

```
def function_name():
    print("My first User-defined Function!")
function_name()
```

Result: My first User-defined Function!

After calling the function, the program jumped back to where the function was defined to run the block of code contained in the body. Once this code executes, the program jumps back to proceed with the code after the function call. You can call this function as many times as you like within the program. It can even be 4 times if you need to run this task that many times:

```
def function_name():
    print("My first User-defined Function!")

function_name()
function_name()
function_name()
function_name()
```

Result: My first User-defined Function!

My first User-defined Function!

My first User-defined Function!

My first User-defined Function!

One thing to keep in mind is that a function has to be defined before actually calling it. You cannot call a function and then define it later in the code. You'll receive an error such as this:

***Result**: NameError: name 'function_name' is not defined*

Okay, so you now know how to create functions. What if you want to add a parameter that can be used in the function? If you create a function that prints out a welcome message, how do you include a user name each time that function is called? All good questions and simple enough to do. Define the function as done earlier. This function is going to be called **welcome_msg**:

```
def welcome_msg():
    print("Welcome to Python 101", name)

welcome_msg()
```

Result: NameError name 'name' is not defined

The body is defined, but the error results from not having a value assigned to the name. We need to create a parameter in our function. This is an easy fix. We just add a name inside of the function's parentheses and type in a name into the call function:

```
def welcome_msg(name):
    print("Welcome to Python 101", name)
welcome_msg("Casey")
```

Result: Welcome to Python 101 Casey

Each time the function is called, it will be looking for input or argument for the parameter name. The function will not run unless a name is supplied! Really cool how we can tailor a function to a single person. Why not just welcome all the students in the course by name?

```
def welcome_msg(name):
    print("Welcome to Python 101", name)
```

```
welcome_msg("Casey")
welcome_msg("Ricardo")
welcome_msg("John")
welcome_msg("Jacky")
welcome_msg("Justin")
```

Result: Welcome to Python 101 Casey

Welcome to Python 101 Ricardo

Welcome to Python 101 John

Welcome to Python 101 Jacky

Welcome to Python 101 Justin

Now alter the code so that there are now two parameters, including the book title and name:

```
def welcome_msg(title, name):
    print("Welcome to " + title + name)

welcome_msg("Python 101", " Casey")
```

Result: Welcome to Python 101 Casey

Using User-defined Functions

On Day 7, you created a calculator. In this calculator, you could add, subtract, multiply, and divide. For any of these operations in an actual program, you might be required to quickly do these operations multiple times.

There is no need to have to repeatedly write the same code over and over again because you can create a function. That function can then be called to perform those operations. Try the code below to see how you can create a multiply function that multiplies two numbers:

```
def multiply(num1, num2):
    result = num1 * num2
    print(result)
multiply(2, 3)
```

Result: 6

With this function, you could quickly multiply several sets of numbers together:

```
def multiply(num1, num2):
    result = num1 * num2
    print(result)

multiply(2, 3)
multiply(4, 2)
multiply(1, 6)
multiply(10, 10)
multiply(3, 5)
```

Result:

6

8

6

100

15

Using what you just learned, you can create a function that calculates the area of a triangle. You might be in a math course that requires you to find the area of various shapes. You can just create a function to do this for you. For now, let's build a function for the area of a triangle:

```
def triangle_area(base, height):
    area = (0.5 * base * height)
    print("The area of the triangle is:", area)
triangle_area(2, 5)
```

Result: *The area of the triangle is: 5.0*

What if I didn't want to print anything out for the function? Usually, a function just executes code and simply returns a result. That result is then normally taken in as an input for executing other operations in the program. In other words, that return result affects the output of other blocks of code within the program.

The first triangle function worked well, but as I said earlier, print statements within functions usually limit your ability to further use the output in the actual program. If I made a function call three times for three different sized triangles and wanted to find the total area, it would be difficult using the previous code. Let's try a new method with just returning a result or area for the function, so the only output returned is a numerical value.

```
def triangle_area(base, height):
    result = (0.5 * base * height)
    return result

triangle1 = triangle_area(2, 5)
print(triangle1)
```

Result: 5.0

The return function just returns the result of the function but doesn't actually display anything. Once the function is called in the second to last line, the return value saves the result as a number in variable triangle1. The result is then printed out using a print statement.

With the output being a numeric variable, it is easier to further process the result later in the program. For instance, you might want to convert this answer from feet to meters. There are 3.281 feet in one meter, so you could take this result which is assumed to be in feet, and divide by 3.281 to convert it to meters:

```
def triangle_area(base, height):
    result = (0.5 * base * height)
    return result

triangle1 = triangle_area(2, 5)
print(triangle1, "feet")
print(triangle1 / 3.281, "meters")
```

Result: 5.0 feet
1.5239256324291375 meters

There you have it. The function output an area of a triangle in feet and converts that number to meters. Now, how about getting the sum of three triangles with different bases and heights in meters?

```
def triangle_area(base, height):
    result = (0.5 * base * height)
    return result

triangle1 = triangle_area(2, 5)
```

```
triangle2 = triangle_area(8, 3)
triangle3 = triangle_area(1, 9)
sum_in_meters = (triangle1 + triangle2 + triangle3)
/ 3.281
print(triangle1, "feet - Triangle 1")
print(triangle2, "feet - Triangle 2")
print(triangle3, "feet - Triangle 3")
print("The sum of all triangles is:", sum_in_meters,
"meters")
```

Result: 5.0 feet - Triangle 1
12.0 feet - Triangle 2
4.5 feet - Triangle 3
The sum of all triangles in meters is: 6.55288021944529 meters

That was not too difficult to do at all. That value, however, is not too pleasant to look at. Thankfully, a built-in function is available to round our result called *round( )*.

```
def triangle_area(base, height):
    result = (0.5 * base * height)
    return result

triangle1 = triangle_area(2, 5)
triangle2 = triangle_area(8, 3)
triangle3 = triangle_area(1, 9)
sum_in_meters = (triangle1 + triangle2 + triangle3)
/ 3.281
print(triangle1, "feet - Triangle 1")
print(triangle2, "feet - Triangle 2")
print(triangle3, "feet - Triangle 3")
print("The sum of all triangles is:",
round(sum_in_meters, 2), "meters")
```

Result: 5.0 feet - Triangle 1
12.0 feet - Triangle 2
4.5 feet - Triangle 3
The sum of all triangles in meters is: 6.55 meters

That is what you were looking for. That sum looks a lot better than a number that goes 10 decimal places. The round function has two parameters that you just filled in. It has the sum and the number of decimal places which is 2. Simply increase that last argument if you want a more precise rounded value.

The critical takeaway is using user-defined functions to use return values when possible. You can now see how hard the sum of all triangles code would have been if you relied on a function with embedded print statements.

* * *

Q&A Review

1.How do you create or define a function in your code?

1. function
2. funct
3. def
4. define

2.How many times are you allowed to use a function in a program?

1. Unlimited

2. Once
3. 3 times
4. Twice

* * *

Day 8 Challenges

- Kilobyte Challenge: Update the triangle_area function you created today to use the built in function multiply.

9

Day 9: Modules & Packages

Reading Time: 10 minutes

Environment Setup

For this chapter, you will need to create an entirely new project called **pythonModules**. This will make it easier to work with several modules and packages. It has been some time since you needed to create a project. If you forgot how simply go to File at the top left of the screen and select New Project... The New Project menu appears. After the last backslash in the location field, replace the text with the name of your new project, **pythonModules**.

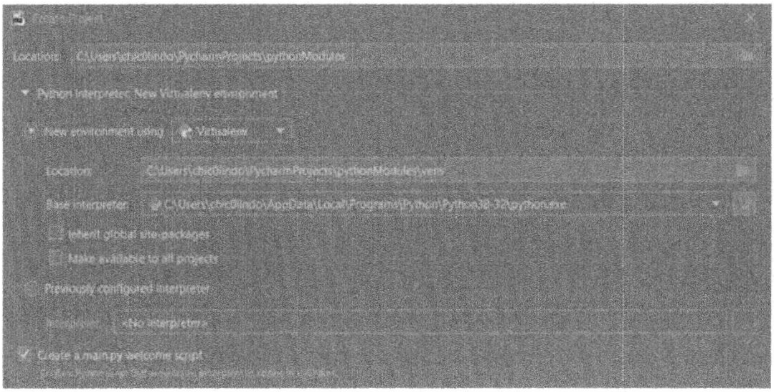

Once create is selected, the environment is set up. At this point, a main.py file is created in your project file list. That main.py file will be used as our program's main file. In other words, it's the file that will hold all the code for executing the program. Let's clear all the default code from your screen.

Modules

Modules consist of Python files containing definitions and statements that can later be used within our program. Modules provide us with the flexibility to organize our program into smaller chunks making it more manageable. This is especially true for those larger programs requiring lots of code. It is sort of the same concept we touched upon in the previous chapter. Think of a module as a library of functions you want to use throughout your program.

The **main.py** file created with your new **pythonModules** project contains the main instructions to execute the program. The modules that you'll be creating will feed into that **main.py** program. That leaves us with the ability to easily reuse code from your modules or even use modules within other modules. At the end of the day, everything feeds into **main.py,** which I'll also refer to as your main program.

Importing Modules

Python has several modules that come built-in to Python 3. Let us take a look at a few of them so you can learn how to import them. For a list of all of the installed modules that come with Python 3, refer to https://docs.python.org/3/py-modindex.html. This list is composed of built-in and external libraries which can be accessed in your Python folder under external libraries. Note that on Windows, you have access to the entire library plus more by default. For Mac and Linux, the use of packaging tools may be required to access some of these libraries.

The three built-in modules I want you to import are os, operator, and math. Head over to the main program or **main.py**. You can import modules anywhere in your program. It is common practice to import at the top of the main program.

The first built-in module for you to import is **os**. The os module is used to perform operating system tasks. The syntax for importing a module is *import* . To import os, you would simply write import os. This would import the entire os module, which includes all of its functions, such as listing all files and folders in a directory, location of the current working directory, and modifying a directory.

For this example, get a list of files in your current directory. You can call a function from a module by typing the period character '.'. Attempt to print out *os.listdir( )*. This is the function used to print out a list of files and folders in a directory. Since you aren't specifying the location of the folder in parentheses, the default value is the current project or working directory:

```
import os
print(os.listdir())
```

Result: ['.idea', 'customModule.py', 'main.py', 'main_alternate.py'...]

Your output shows all the files in your current working directory. Mine differs from yours by a few extra files. That is okay; you get the idea! One thing you might be wondering is what directory you are currently working in. One of the parameters for the **listdir** function is specifying the location of a directory. You can use another function to retrieve the location of your current working directory:

```
import os
print(os.getcwd())
```

Result: E:\dev-projects\course-python101

There goes my current working directory. There are many more functions found within **os**. You can view all of them on the Python 3 documentation website, and I provided the link earlier in the chapter.

In chapter 9, I defined a few math functions. Little did I know there is already a module that contains some of those same operators. The name of that module is **operator,** and it contains all the standard operators as functions. Import the **operator** module below the **os** module at the top of your code.

Now you can use an operator to perform all those standard math functions without having to define new functions:

```
import os
import operator

print(os.getcwd())
print(operator.add(5, 6))
print(operator.sub(5, 6))
print(operator.mul(5, 6))
print(operator.truediv(5, 6))
```

Result: *E:\dev-projects\course-python101*

11

-1

30

0.8333333333333334

The operator module allowed us to quickly do common math calculations on the numbers **5** and **6**. Did I mention there is also a module called math!? The math module has a giant list of math functions ranging from simple math calculations to more advanced. We will take a look at 4 of the many functions within the math module:

- **math.pow(x, y)** - returns x raised to the y power
- **math.sqrt(x)** - returns the square root of x
- **math.fabs(x)** - returns the absolute value of x
- **math.gcd(x, y)** - returns greatest common divisor of integers x and y

Like you did before, it's time to import the math module into your code. You will add the above 4 functions to the rest of your code with random numbers to see what you get:

```
import os
import operator
import math

print(os.getcwd())
print(operator.add(5, 6))
print(operator.sub(5, 6))
print(operator.mul(5, 6))
print(operator.truediv(5, 6))

print(math.pow(5, 6))
print(math.sqrt(25))
print(math.fabs(-8))
print(math.gcd(48, 36))
```

Result: E:\dev-projects\course-python101

11

-1

30

0.8333333333333334

15625.0

5.0

8.0

12

So far, you have been importing entire modules. This is not ideal if you only plan on using only one function in a module. To import just a function in a module, use the syntax *from import*. Say you want to use only the add function from the **operator** module. Instead of importing the entire **operator** module, you would write **from operator import add** at the top of your code. Then you can perform the same add function as before:

```
from operator import add
print(add(5,6))
```

Result: 11

With the code above, you imported the add function. Also, note that you didn't have to specify the module when calling the add function either. Okay, let's revert back to importing the entire module:

```
import operator
print(operator.add(5,6))
```

Result: 11

The reason for reverting back is I wanted to show you how you could rename the module and use the new name to call the function. Okay, instead of operator, let's rename the **operator** module to **ops**. To do that, simply write **import operator as ops**. Now you'll use ops to call the function:

```
import operator as ops
print(ops.add(5,6))
```

Result: 11

You receive the same result after you rename your module. The only difference is you can use the new name to call the operator's functions. One last thing I forgot to mention is how to find out what functions are defined within a module. The built-in function *dir()* can be used to list all functions in a module:

```
import math
functions = dir(math)
print(functions)
```

Result: ['doc', 'loader', 'name', 'package',' spec'...]

The output displays the list in alphabetical order. As you can see, the math module has a ton of functions defined. If you would like an easier way to look at this list, consider using a for loop:

```
import math
functions = dir(math)
for function in functions:
    print(function)
```

Result: ['acos', 'asin', 'ceil'...]

Creating Modules

I already discussed how to import pre-existing modules. Not every module out there, however, fits the use case for your program. If you'd like to create your own modules, that can be done as well.

Right-click your project folder on the left in the project files pane, then click on New. Give your file the name **customModule** and select the Python file option. This creates a file with the name **customModule.py**. The suffix .py indicates that it is a module. Without the .py suffix, you would not be allowed to import the module into your main program.

Open up your new module. Type in the following code that prints thank you to a user for using the custom module:

```
def thanks(name):
    print("Thank you " + name + " for using this new
    module.")
```

If you run this module in the current state, nothing is displayed because the function **thanks** have yet to be called. Now navigate back over to your main.py file. Clear your screen with any leftover code from the previous section.

Your custom module can now be imported like any other module. Import the module, then call the function with a name as a parameter:

```
import customModule

customModule.thanks("Casey")
```

Result: Thank you Casey for using this new module.

Look at that. You were able to import and use the module you created. That was really neat.

Python Community Modules

Python is open-source with a community of developers and supporters. This community of Python developers provides software and libraries for everyone to use. Chances are, if you are trying to accomplish a task in a program, someone has already done it. The community is great because you can leverage code that has been written by another developer. There is a myriad of non-standard modules available for use in your programs. Python Package Index (PyPI) is the default repository where community Python packages can be found. Try searching the https://pypi.org/ website to look at some of the available packages.

So, what exactly is a package? A **package** consists of all the necessary files required to use a module. You can install packages from the Python package repository PyPI. In order to do this, you have to ensure you have the package-management system known as pip installed. Pip is the default Python package manager used to install, update, uninstall, and manage software packages.

Python pip (Package Manager)

pip comes installed on every version of Python after version 3.4. You can check to see if pip is installed on your system. If you are running Microsoft Windows, go to the Start Menu, type in CMD, and then hit enter on the keyboard. A command prompt should appear. In the command prompt, type in **pip –version** and hit enter:

As you can see, I am running version 21.1.2 of pip, and I'm also given the path where it is installed.

For Mac/Linux/Unix users, open up a terminal. Type in **python -m pip –version,** then enter. This will show you the version of pip that is currently installed on your system.

If pip is not installed on your system, navigate over to https://pip.pypa.io/en/stable/installing/. There are directions on how to install pip on the appropriate operating system. Chances are, at the time of reading this book, pip is already installed on your system.

Installing Community Modules

After verifying that your package manager pip has been installed, you can now install community, 3rd party, or external modules. There are tons of packages to choose from out there. The one thing I haven't talked too much about in this course is a Graphical User Interface or GUI for short. Most applications use a GUI to allow users to interact with the program easily. With that said, let's install a GUI module that will allow our programs to react like a more modern-day application.

The package you are going to install is **PySimpleGUI**. In short, PySimpleGUI gives us the ability to add graphics to your programs! Most people want some type of GUI in their programs to make it more intuitive.

All right, let's get on with installing PySimpleGUI. I hope you are as excited as I am. Open up your command prompt or terminal again. If you are using Windows, type in **pip install PySimpleGUI** followed by enter. For Mac/Linux/Unix users, input the command **python3 -m pip install PySimpleGUI** into your terminal followed by enter. After running the command, it will start installing the PySimpleGUI package for you. The last line should let you know that you have successfully installed the latest version of PySimpleGUI.

```
C:\Users\chic0lindo>pip install PySimpleGUI
Collecting PySimpleGUI
  Downloading PySimpleGUI-4.43.0-py3-none-any.whl (350 kB)
     |████████████████████████████████| 350 kB 1.7 MB/s
Installing collected packages: PySimpleGUI
Successfully installed PySimpleGUI-4.43.0
```

It's just that simple to download and install packages. In your command prompt, you can also enter the command **pip freeze** to display all packages you've installed. Search through https://pypi.org/ for any other packages you'd like to install.

Using PySimpleGUI package

Create another python file called **my_GUI.py**. Import the **PySimpleGUI** module. I'm going to do an **import PySimpleGUI as sg**. So each time I call a function from PySimpleGUI, the only thing I have to put prior to the function is **sg**. It just keeps me from having to spell out PySimpleGUI each time I use the module. We will be using sample code from the developer. Copy and paste the code below into your my_GUI.py window:

```
import PySimpleGUI as sg  # Part 1 - The import

# Define the window's contents
layout = [[sg.Text("What's your name?")],  # Part 2
- The Layout
          [sg.Input()],
          [sg.Button('Ok')]]

# Create the window
window = sg.Window('Window Title', layout)  # Part 3
- Window Definition

# Display and interact with the Window
event, values = window.read()  # Part 4 - Event loop
or Window.read call

# Do something with the information gathered
print('Hello', values[0], "! Thanks for trying
PySimpleGUI")

# Finish up by removing from the screen
window.close()  # Part 5 - Close the Window
```

Result:

Run this code and see what happens. If everything checked out, you should've seen a GUI pop up prompting you for your name. After entering your name and selecting Okay, the window disappears, and an output on your run screen displays a thank you for reading this book. The sample code provided above contains just some of the functionality you now have available within this module. The User's Manual on the website provides step by step instructions on how to use all the features of the PySimpleGUI package https://pysimplegui.readthedocs.io/en/latest/

* * *

Q&A Review

1.What keyword do you use to use a package or module in your Python code

1. download
2. use
3. import
4. run

2.What is the package installer / manager for Python?

1. pip
2. pyp
3. snake
4. cobra kai

* * *

Day 9 Challenges

- Kilobyte Challenge: Find a 3rd party package on pypi.org that seems interesting and try it out in a program. See https://pypi.org/search/ - If you can't find one that looks cool check out Colorama - https://pypi.org/project/colorama/ :)

10

Day 10: Working with Files

Reading Time: 11 minutes

In Python, your applications will often need to pull or push data from external files. These files do not necessarily need to be python files. They can be any type of file, ranging from a text file to an executable. Different supported file types are represented by their file extensions (e.g. .html, .txt, .csv, .bin, .sh). Python has file handling capabilities that allow a program to read, write, create, and/or append to external files. Today, you will use standard text files with the extension **.txt**.

To open any file in Python, use the built-in function *open( )*. The function accepts two arguments within the parentheses: the file you'd like to open and the mode you'd like to open the file in. Only the *<file_name>* is needed as a parameter for files located in the same working directory. The file location must also be included for files located outside the working folder or directory.

Let's learn how to open a file. The only parameter required to open a file is the file name. If the second parameter for mode is not provided, the *open( )* function will default to reading the open file. In the left window pane in your project files, right-click on your project to create a new file. Name this file **nerd_names.txt**. This creates a blank text file in the current working directory, which you can double click to edit. Copy and paste the following in your **nerd_names.txt** file:

```
Nikola
Tesla
Einstein
Thomas
Edison
Ricardo
Casey
Kelvin
```

Create a new python file in the same project called **readFile.py**. This will represent your program to open and read in the text file you created. In the **readFile.py** code window, use the *open( )* function to open the new text file. The syntax for opening the new file is **open(nerd_names.txt)**. Since the mode parameter is omitted, the file will be opened and read. Every file opened using this function must also have a closing function. The closing function is written after all operations on the file have been completed. It's good practice to write the closing function when you write the open function for a file. Take a look at how this is done below:

```
nerd_file = open("nerd_names.txt")
nerd_file.close()
```

Result: Exit process with code 0

Read Files

The file was opened, then read, and immediately closed. Notice that nothing appeared in the output window. You have not yet told the readFile program to do anything with the file. You have to tell the program what to do with this data. To ensure the program can properly read the data, you can use another function called *readable( )*. This tells us whether the **nerd_names.txt** is actually able to be read by the program:

```
# a "True" output indicates program can read file
nerd_file = open("nerd_names.txt")
print(nerd_file.readable())
nerd_file.close()
```

Result: True

The printed response **TRUE** indicates that the file is readable by the program. A response of **FALSE** would have indicated that the file is not readable by the program. This probably would result from a file outside the current working directory, misspelled name, or a corrupted file. A file outside the current working directory would just need a path to specify its location. I will talk more about this later on. Okay, now print out a copy of the entire text file in the program using the *readlines( )* function:

```
# prints out all lines in an array
nerd_file = open("nerd_names.txt")
```

```
print(nerd_file.readlines())
nerd_file.close()
```

Result: ['Nikola', 'Tesla', 'Einstein', 'Thomas', 'Edison', 'Ricardo', 'Casey']

The output of readlines prints out all the lines of the text file as a list. Remember that the is an escape character that is just showing a new line. The readlines function is printing out each line of the text file. If you just wanted the first line, you could use the singular function *readline( )*:

```
# reads and prints the first line in nerd_file
nerd_file = open("nerd_names.txt")
print(nerd_file.readline())
nerd_file.close()
```

Result: Nikola

If you use call the *readline( )* function twice it will output the first two items of the text file:

```
# reads and prints first 2 lines in nerd_file
nerd_file = open("nerd_names.txt")
print(nerd_file.readline())
print(nerd_file.readline())
nerd_file.close()
```

Result: *Nikola*

Tesla

Okay, these are some excellent functions. Now, what happens when you want to choose an item to print within the text document? Say you want to select the third item within the document, which in this case is **"Einstein"**. How can you read the file and print out this single item? Since **readlines()** give you a list, you can use indexing. The third item is also index 2. Let's print it out:

```
# reads and prints out 3rd line or item in the file
nerd_file = open("nerd_names.txt")
print(nerd_file.readlines()[2])
nerd_file.close()
```

Result: *Einstein*

Okay how about we keep it simple and print out exactly what is in the text file using the *read( )* function:

```
nerd_file = open("nerd_names.txt")
print(nerd_file.read())
nerd_file.close()
```

Result: Nikola

Tesla

Einstein

Thomas

Edison

Ricardo

Casey

Reading Files Outside Directory

Chances are a file you need to open won't be in the same working directory of where your script runs from, so how do you specify a directory? Reading a file outside the current working directory requires adding a path before the filename. Go to your desktop of your PC, Mac, or Linux and add a file called **testPythonfile.txt**. Also, copy down the exact path of this file (e.g., C:/Users/Username/testPythonfile.txt). Open the file and input the following text:

```
Circle
Rectangle
Triangle
Square
```

Save the file once you are done, then close the window. Alter your original code in **readFile.py** with the following and use your actual folder location, not mine:

```
desktop_file =
open("C:/Users/Rico/Desktop/testPythonfile.txt")
print(desktop_file.read())
desktop_file.close()
```

Result: Circle

Rectangle

Triangle

Square

You were able to open a file outside of the current working directory. It's essential to know the exact path of the file you are accessing and to insert that path before the filename in the function.

Access Modes

Remember the *open( )* function has more than one parameter. It has a second optional parameter we have yet to discuss called access modes. The default mode is read, which is represented by the letter *r*. There are a total of six modes, but here are the 4 I will be going over in this section:

- **r** - Open specified file to read
- **a** - Open specified file to append
- **w** - Open specified file to write
- **x** - Create a specified file

The *open( )* function uses read as the default value. So **nerd_file = open("nerd_names.txt")** is equivalent to **nerd_file = open("nerd_names.txt", "r")**. To use any other modes, simply replace the second parameter or argument with the corresponding mode letter. The next sections go into examples of this.

Append to Files

Create another text file in the current working directory called **more_names.txt**. In the text file, put the following names:

Newton
Hawkings
Tyson
Musk

This time create a python file with the name **appendFile.py**. Let's do a quick check to make sure that the more_names.txt file is readable:

```
names_file = open("more_names.txt", "r")
print(names_file.readable())
names_file.close()
```

Result: True

As you learned earlier, **TRUE** means that the file is readable by your program; since it is readable, you can access the file to append more content to it. When you use append, you can add text and even contents from another file to the end of the read file. So, information from one file can be appended to another, resulting in that one file having the contents added from the appended file. We use the *a* as the mode for the *open( )* function to append a file. The syntax is *open(filename, "a")*.

It is essential to understand that you are not just opening and reading a file when using this mode. You are also writing to it. The append function will write to the end of the file you open to add additional contents. There is no reverting back to the original file after the function has been executed. It is a good idea to back up your original file in case a mistake is made.

All right, well, let's give it a go. In our example, you will append the name **"Gates"** to **more_names.txt**. Then you will print out **more_names.txt** to see how it looks:

```
names_file = open("more_names.txt", "a")
names_file.write("\nGates")

names_file = open("more_names.txt", "r")
print(names_file.read())
names_file.close()
```

Result: Newton

Hawkings

Tyson

Musk

Gates

As you can see, you were able to add **"Gates"** to your **more_names.txt** file. I added the escape character *** so that the name would be on its own line like all the other names already in the file. If you don't add that escape character, the name would be added right after the last string in the file; in this case, it would be Musk. Without** *, your code would output the last line as **"MuskGates"**.

Let's take that same example. Using that same code from before, change the file in both open functions to **"random.txt"**. You and I know that there is no file in the current working directory with that name. I wonder what is going to happen. Only one way to find out:

```
names_file = open("random.txt", "a")
names_file.write("\nGates")
names_file.close()

names_file = open("random.txt", "r")
print(names_file.read())
```

```
names_file.close()
```

Result: Gates

There goes the output only with the Gates name. Your code executes and tries to open the file by looking in the current working directory. Since the file cannot be found, it creates a file with the corresponding name **"random.txt"**. There is no content in the new file that was created, so only the name **"Gates"** is visible, as it has been appended to a blank file. If you were to rerun this program, **"Gates"** would be added again to another line, so be careful.

Append One File to Another

You can append the contents of one file to another file. Both files must be opened and later closed. For practice, we are going to append **nerd_names.txt** to **more_names.txt**. This means that the contents of **nerd_names.txt** will be copied over to the end of the **more_names.txt** file.

```
# Opens both files, including one to be appended to
and the other to read
names_file = open("more_names.txt", "a")
nerd_file = open("nerd_names.txt", "r")

# Appends the nerd_names.txt content to
more_name.txt file
names_file.write(nerd_file.read())

names_file = open("more_names.txt", "r")
print(names_file.read())
```

```
names_file.close()
nerd_file.close()
```

Result: Newton

Hawkings

Tyson

Musk

Gates

Nikola

Tesla

Einstein

Thomas

Edison

Ricardo

Casey

As you can see, you now have a list containing the original contents of **more_names.txt** with the **nerd_names.txt** contents added to it.

Write

Create another file called **writeFile.py**.

Using the write mode or *w* writes to the file being opened. It will overwrite all the contents already in the file. So, if you just want to add to a file, ensure your mode is set to *a* and not *w*. If not, all contents will be replaced with the information in the write function.

Okay, you made the decision that writing to a file is needed.

You are not going to create the file beforehand. Like you did with append, if the open function doesn't see a file when called, it creates it. Your objective is going to be for your program to generate a text file with a list of vowels:

```
# Creates a new file ready to write to.
vowels_file = open("vowels_list.txt", "w")

# Creates a list of vowels and writes list as a
string to new text file
vowels_list = ["a", "e", "i", "o", "u"]
vowels_file.write("This is a list of all the
vowels:" + str(vowels_list))

# Opens and prints out updated vowels_list file
vowels_file = open("vowels_list.txt", "r")
print(vowels_file.read())

vowels_file.close()
```

Result: This is a list of all the vowels: ['a', 'e', 'i', 'o', 'u']

The code above created a new file called **vowels_list.txt** and wrote the outputted contents to it. In doing this, you created a list of vowels into a string so it could be written to the file. Remember the parameters for the *write( )* function are strings. While the list contains a string of vowels, it still isn't a string. This is why you had to convert the list into a string using explicit type conversion.

Create Files

With the append and write functions, you were able to create new files. There is another way to just call the open function to create a new file with no contents. First, create a new python file called **createFile.py**. The syntax for creating a new file is *open"file", "x")*. The *x* mode is used to create a new file. It can be any file name with an extension added on to the end. Try creating a new file of your choice:

```
new_file = open("created_file.txt", "x")
new_file.close()
```

There isn't any visual output when executing the associated code. However, if I open my current working directory, a new blank file called created_file.txt in my case appears. Find the name of the file you created with the name you chose.

Delete Files

It was fairly simple to create files. In Python, you can also delete files and folders. Just as you can create files, you can easily delete files. To delete files, you must import the **os** module. The function for removing files is *remove( )*. Create a new python file called **removeFile.py**.

Let's remove the last file you created using the remove function:

```
# Deletes the file named created_file.txt from the
working directory
```

```
import os
os.remove("created_file.txt")
```

This code removed the file you created from the current working directory. You can also remove files from other directories like your desktop. Remember that file you created earlier **testPythonfile.txt**. You can delete that file by including its path:

```
# Deletes the file on my desktop named
testPythonfile.txt
import os
os.remove("C:/Users/Rico/Desktop/testPythonfile.txt")
```

Delete Folders/Directories

Folders or directories can be deleted in the same manner as files. The only difference is the function used for directories is *rmdir( )* instead of *remove( )*. The *rmdir( )* function can only delete empty folders. An error will occur if you try to remove a directory with files. You must first delete all the files in a folder or directory before removing the directory using *rmdir( )*. "rmdir" is short for remove directory. The *remove( )* function does not allow you to remove a directory.

Right-click on your **pythonModules** project folder, then select New then directory. Name the directory **testDirectory**. Now create another python file named **removeDirectory.py**. In this file, enter the following code to permanently delete the directory using the *rmdir( )* function:

```
# Deletes the folder named testDirectory from the
current working directory
import os
os.rmdir("testDirectory")
```

After executing that code, your program removed the **testDirectory** that was just created. This function allows you to remove any empty folder.

* * *

Q&A Review

1.Using the open and write function for an existing file:

1. will append contents to the file
2. will overwrite all the contents already in that file
3. will create a file with the same name and append contents to that file
4. will delete all contents within the file so it's empty.

2.The following code block will open a file called nerd_names.txt, read it, and print it to the terminal and then subsequently close the file.

```
nerd_file = open("nerd_names.txt", "r")
print(nerd_file.read())
nerd_file.close()
```

1. True
2. False

* * *

Day 10 Challenges

- Kilobyte Challenge: Use python to write an array of your favorite meals to a text file.

11

Day 11: Debugging

Reading Time: 8 minutes

Hello Nerds! It's day 11, and for today's lesson, you will be learning about debugging. The Python debugger is an interactive code tool that allows you to debug your code. You can evaluate variables, step through the code line by line, and set breakpoints. You're going to learn about the Pycharm interface and some abilities it has through the Python debugger. If you don't have PyCharm installed, any modern IDE will offer these capabilities; however, the shortcodes will likely not be the same. At the end of today, you will solve your second Nerd Challenge! Let's jump right in!

Breakpoints

When using the Python debugger, a breakpoint allows you to stop the execution of your program at a particular line. You can do this in Pycharm using the shortcut Ctrl+F8 on your keyboard, but first, you need to have some code to add a breakpoint. Write the following code in your PyCharm IDE.

```
debugger = "A debugger is a tool that lets you find
problems or bugs with your code."
print("Welcome to this fake program")
print(f"We will use it to talk about a debugger.")
print(f"{debugger}")

add = 5 + 5
print(add)

def some_cool_function(x, y):
    result = x * y
    return result

run_this = some_cool_function(3, 3)
print(f"{run_this}")

user_x = int(input('Enter a number you want to times
by 2 > '))
user_result = some_cool_function(user_x, 2)
print(f"{user_result}")
```

Place a breakpoint on the line: ‘add = 5 + 5‘ line. To place a breakpoint, you can select the line and press Ctrl+F8. Alternatively, you can click on Run > Toggle breakpoint> Line Breakpoint. You added the breakpoint, and you can now see the red dot on line 6. The red dot is the icon for breakpoints in PyCharm.

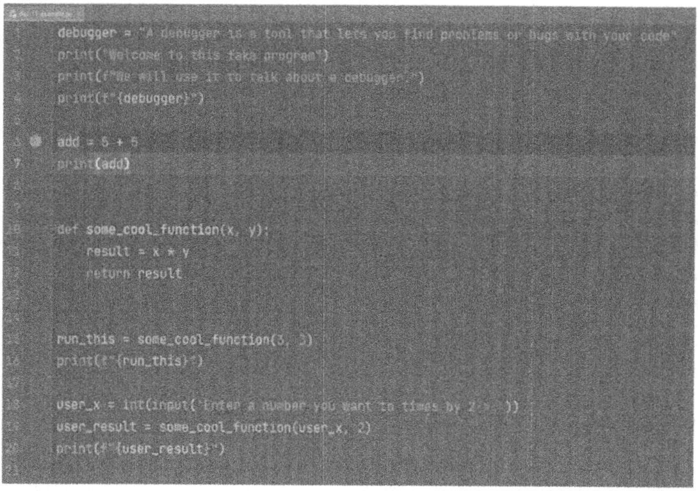

Running the debugger

You have placed the breakpoint; next, you need to run the debugger. From the menu, choose Run > Debug. Alternatively, you can use Alt+Shift+F9 on your keyboard. PyCharm may prompt you with which file you would like to use. Select the appropriate file. In my case, it is 'example.py'. You can now see a debugger window on the bottom half of your screen.

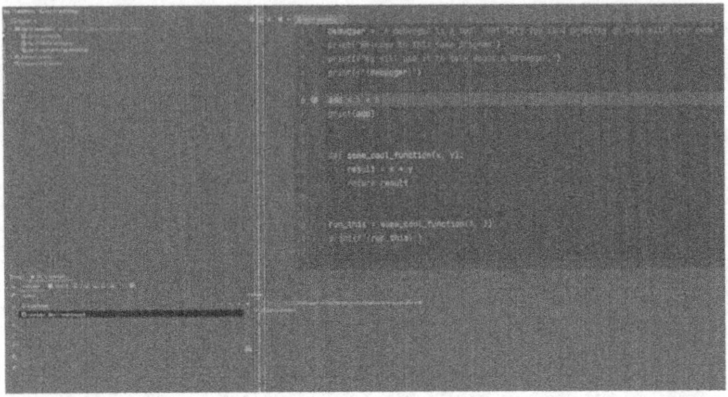

You can see that the program has stopped execution at line 6, and you can view the output via the 'Console' window. Click on the 'Console' tab, and you can see the following result:

Welcome to this fake program

We will use it to talk about a debugger.

A debugger is a tool that lets you find problems or bugs with your code

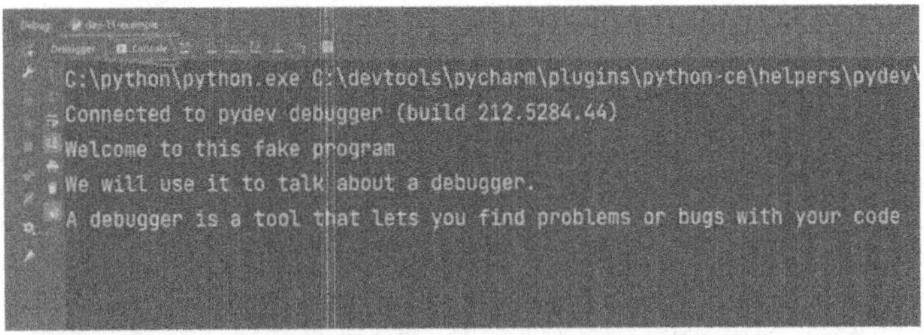

As you can see, the debugger has performed all of the code up to, but excluding line 6. Python assigned the 'debugger' variable and printed out the three statements you specified. Click back on the 'Debugger' tab, and we'll discuss a few more options.

Watchers

A watcher allows you to evaluate any number of variables or expressions in the current execution, specifically in something PyCharm calls a stack frame. I won't be covering stack frames or CPU threads in this book as it is an advanced topic. Still, you'll find more information in the resources section if you're interested in learning more. For now, it's essential to know that a watcher gives you the ability to keep an eye on or *watch* a variable or expression as you step through the application. If you specify a watcher, you can monitor how it changes over time as your program runs.

Adding a Watcher

You know what a *watcher* does, so it is time to add one. Click on the *Debugger* tab and then in the Variables window, you see a + sign and create a watcher for 'add', which is the variable you've defined on line six. PyCharm will report an error. The error is

{NameError} name 'add' is not defined

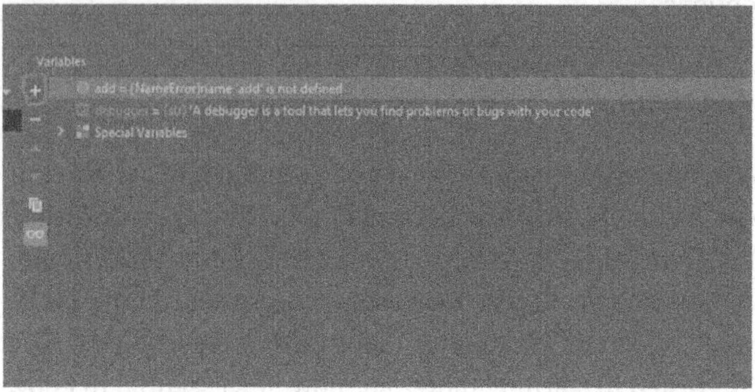

Why do you think this is occurring? The breakpoint stops at line 6. It does not execute the line and then stop; therefore, add does not exist from Python's perspective. You will learn how to continue Python's execution of your program in the next section. Also, notice that the *debugger* variable automatically shows up as a watcher because PyCharm automatically adds variables that it finds as a *watcher*. So, you didn't even need to add the 'add' variable. It would have happened automatically!

Stepping through your code

To step through code means to execute one line of code at a time until you finish the program or get to a point where you want to stop. You're going to go through your code line by line to help you identify any problems or spot any bugs. PyCharm offers five key options for *stepping* through your code: Step Over (F8), Step Into (F7), Step Into My Code (Alt+Shift+F7), Step Out (Shift+F8), and Run To Cursor (Alt+F9).

You're going to learn about each of these options in the following sections; however, after reading about each option, I want you to perform each of them using the example code.

By using the debugger, you gain a better understanding of how *stepping* through code works. If you finish executing all of the lines of code in the example program, you can always rerun the debugger. After all, you know how to do that now! One **important** note, when you get to line 18, the program asks for *input()* from the user, so when you hit this line, you'll need to go to the Console tab and enter in a number you want to times by 2. Then the debugger will continue the execution of the program.

Step Over

Step over will step over the current code line and take you to the following line of code. The vital characteristic here is that it will go to the following line of code even if there is a function or method call on that line. In the example code, you have a function named '*some_cool_function*', which you call on line 15. If you were to step over on line 15, Python would run that line without going into the definition of that function you defined on lines 10 through 12. Try this out so you can see for yourself!

Step Into

Step Into is the opposite of *step over*. Step into means that if any functions are defined on the current line that the debugger is evaluating, the debugger will *step into* those functions. Try this functionality out in the example code you set up; what happens now when you call the 'some_cool_function' on line 15? Where did 'step into' take you? Why?

The step into feature will take you into the int() and input() functions, which are core Python functions. It is possible, but unlikely, to have errors or bugs in these functions, so you may not want to specifically step into them. You can use the step-over operation for those specific lines to get around this. Alternatively, PyCharm gives you another option that alleviates this problem.

Step Into My Code

That's right! You can step into only your code, which is one of my favorite features of PyCharm. You can step through your code and ignore any libraries you are calling. This is very handy if you don't suspect any issues with other libraries you are using! Try it out on your example code to see it in action.

Step Out

If you step into a function and perhaps you didn't mean to, or you want to check something. Still, you confirm that there isn't a problem with that function. PyCharm gives you the ability to step out of it with this option. Rerun the code using step Into, and after you hit line 18, use step out to jump out of *parse.py*. Awesome right?

Stopping the debugger

Sometimes, you may want to stop the debugger. Perhaps you've looked through a portion of the code and found the error, or maybe the debugger gets stuck, and you need to exit. To quit the debugger, you can press the shortcut key (Windows) 'Ctrl+F2' or go to 'Run > Stop file-name'.

* * *

Q&A Review

1.A breakpoint allows you to

1. automatically breaks your code
2. allows you to stop your program at a specific line number and then step through subsequent line codes
3. only works on lines with integers on them
4. only works if a variable has been defined

2.A watcher allows you to watch a variable as your program runs.

1. True
2. False

* * *

Nerd Challenge 2: Debugging a file

You've learned an extensive amount about Python in just 11 days, and you're growing your skills and continuing to improve them. Just as a quick recap, here's what you've learned so far:

- **Day 1** – you installed Python and were introduced to variables

- **Day 2** – you learned about data types and intro to basic operators
- **Day 3** – you dove into more advanced operators
- **Day 4** – you learned how to prompt users for input
- **Day 5** – you know how to create if-else statements and loops
- **Day 6** – you built a program that handled exceptions and errors
- **Day 7** – you solved your first Nerd Challenge, where you made a simple calculator
- **Day 8** – you programmed your first defined Python function
- **Day 9** – you imported your first third party package and created your own
- **Day 10** – you learned how to manipulate files with Python

Congratulations to you for completing all of this in just ten days! This challenge is a bit different; I have written the code for you. Your job for this challenge is to identify the primary bugs that exist within this python application. As you find problems, note them on a piece of paper or in a separate file, try and fix the problems yourself, and then compare your answers with the solution. You may find different bugs than the ones we've listed in the Solution section, which is perfectly fine; it isn't an exhaustive list.

So are you ready for the challenge? If so, you know the drill, say **Challenge Accepted!**

Buggy File: 'day-11-nerd-challenge.py'

```
# Nerd Challenge: Debugging
# Erm, I found some bugs :(
import random

guess = input('Guess a number between 1 and 5: ')
x = random.randint(0, 5)

print(f"You guessed {guess}")
print(f"The answer was: {x}")

if guess == x:
    print("Congrats you guessed right!")
else:
    print("Sorry you guessed wrong. You lose.")

random_fact_result = guess / x
print(f"Random fact of the day: Your guess number,
{guess} divided by the answer {x} is equal to
{random_fact_result}")
```

Solution

No peeking! Check the answer once you've spent at least 1 hour on the challenge.

* * *

Solution File: 'day-11-nerd-challenge-solution.py'

```
# Nerd Challenge: Debugging
# Erm, I found some bugs :(
import random

guess = input('Guess a number between 1 and 5: ')
if guess.isdigit() and 5 >= int(guess) >= 1:
    guess = int(guess)
else:
    print("Sorry, next time enter a number between 1
    and 5")
    exit(-1)

x = random.randint(1, 5)

print(f"You guessed {guess}")
print(f"The answer was: {x}")

if guess == x:
    print("Congrats you guessed right!")
else:
    print("Sorry you guessed wrong. You lose.")

random_fact_result = guess / x
print(f"Random fact of the day: Your guess number,
{guess} divided by the answer {x} is equal to
{random_fact_result}")

# Error #1: input returns a string. We need to
compare integers!
# Error #2: The instructions say between 1 and 5,
but 0 is possible
# Error #3: ZeroDivisionError Exception
```

```
# Error #4: TypeError: unsupported operand type(s)
for /: 'str' and 'int'
```

It's been a long but fun and exciting day. You've learned quite a bit about utilizing the debugger, and on day twelve, you will learn about classes and objects in Python.

12

Day 12: Classes and Objects

Reading Time: 8 minutes

Python Classes

Think of **classes** in Python as a blueprint to create objects that can be used throughout your program. A blueprint or schematic for a building provides you with the necessary framework to construct a building. The same goes for classes as it relates to objects. Classes essentially offer you the ability to create your own data type. This is important because not everything in the real world can be represented by the data types I discussed on Day 3. Those data types were limited to a string, number, list, tuple, and dictionary.

Think about some of the people, places, or things you might need to represent in a program. There probably isn't any variable that can store that using the basic data types you learned about on Day 2. This could be a problem if you didn't have classes because you would need to repeatedly generate a lot of code to represent that data type each time you wanted to use it. Classes allow us to create objects, so you don't need to do that.

Another trait of Python that I have yet to discuss is that it is also an object-oriented programming language. Like Java, C++, and C#, it allows organizing a program design around classes and objects instead of solely procedural.

Classes consist of data attributes and methods. The attributes describe the characteristics of the class. It's defining what that class is. Then there are **methods**, which describe what a class does or how it functions. This will become clearer as I go over a few examples.

Objects

Objects are instances of classes. So, what you do is create one class. Each time that class is called, it creates an object or instance from that class. This means that the object created holds all the same attribute types and methods as the class. It also means you can create multiple objects of the same class with a different attribute value.

Constructing Classes

I stated that classes are composed of data attributes and methods. Cars are a perfect example of a class. There are Cars everywhere with different years, makes, models, colors, and doors. All those characteristics I just listed are data attributes. The values of those attributes may be different, but they are always on each car. They are what makes a car a car! Each car also has a few functions that they perform, such as accelerating, braking, and turning. In a class, this would be known as their methods. Each different type of car you represent in your program will be a new object.

Let's create a new Python file called **car_class.py** to show how classes are constructed. The first line in the code will have the class name represented by the syntax *class.* In this case, you'll write **class Car:** to indicate a Car class.

Below your class name will be your ***init( )*** function. This should be included in each function you create. This function always gets executed when the function is initialized. It is the first part of the code that is run when the class is called. Within the parentheses of the function, you can define parameters for the data attributes. The first parameter in an ***init()*** function is always *self.* Self gives us access to the data attributes and methods within the class. Self is then followed by the data attributes of the class:

```
def __init__(self, parameter1, parameter2,
parameter3, etc..):
```

Your data attribute types will be the year, make, model, color, and doors in your Car class. These are the attributes of a car that may have different values. This will be used to create an object out of the class. Let's fill in some of this information for your car class. Instead of doors, just do is_Sedan. If Sedan is **TRUE,** it's a four-door; if **FALSE,** it's a coupe or two doors:

```
class Car:
    def __init__(self, year, make, model, color,
    is_Sedan):
```

The **init** function is not yet finished. You still need to add a body of code that allows the input from the parameters to be saved to the object's variables. Simply stated, when an object is created, the values of its attributes need to be stored. Add the following code below to your object definition to do this:

```
class Car:
    def __init__(self, year, make, model, color,
    is_Sedan):
        self.year = year
        self.make = make
        self.mode = model
        self.color = color
        self.is_Sedan = is_Sedan
```

Create a new python file called **car_program.py**. This is where you'll create instances of the Car class or objects. Type in **from car_class import Car**. This now makes the Car class available for use in your car program. The command imports the Car class from the file you created named **car_class.py**.

Now you're going to create a variable **car_A** and assign a car object to it. You want this object to be a 2021 Tesla Model S. Since you have already created a car class, this is easy. You can call the class and enter in car specific attributes to create the Tesla car object:

```
car_A = Car("2021", "Tesla", "Model S", "Red", True)
```

The object created is now saved in variable **car_A**. You were able to represent an entire car by creating an instance of the class Car. Remember those object attributes you created in the car Class? You can now access those variables since they have been stored in the object. Let's see how this works by printing some of the values of the attributes you created:

```
from car_class import Car
car_A = Car("2021", "Tesla", "Model S", "Red", True)
print("The color of the car is:", car_A.color)
print("Is this car a Sedan, True or False?",
car_A.is_Sedan)
```

Result: *The color of the car is: Red*

Is this car a Sedan, True or False? True

The output was able to print out the car color and door attributes. The beauty of this is that if you had 100 cars and wanted to know one characteristic of a few cars, you could easily print it out. You wouldn't need to know any other identifying information except for the object name.

In this example, you only created one car object. I thought classes were supposed to make it easy to create multiple objects just like you have done in the past with standard data types? You can, and it's simple as creating more Car objects and assigning those to more variables:

```
from car_class import Car

car_A = Car("2021", "Tesla", "Model S", "Red", True)
car_B = Car("2021", "Mazda", "CX-5", "White", True)
car_C = Car("2014", "Cadillac", "CTS", "Black",
False)

color_list = [car_A.color, car_B.color, car_C.color]
door_list = [car_A.is_Sedan, car_B.is_Sedan,
car_C.is_Sedan]
```

```
print("The colors of those cars are:", color_list)
print("Are those cars a Sedan, True or False?",
door_list)
```

Result: *The colors of those cars are: ['Red', 'White', 'Black']*

Are those cars a Sedan, True or False? [True, True, False]

You were able to create multiple Cars and print out color and door attributes for each using lists. If you wanted to get fancy, you could loop through each list to find the number of colors that pop up in inventory in your inventory of cars. Let's change the color of your car A Tesla object to **"Black"**. Then you'll add some additional code to print out the number of Black cars:

```
from car_class import Car

car_A = Car("2021", "Tesla", "Model S", "Black",
True)
car_B = Car("2021", "Mazda", "CX-5", "White", True)
car_C = Car("2014", "Cadillac", "CTS", "Black",
False)

color_list = [car_A.color, car_B.color, car_C.color]
door_list = [car_A.is_Sedan, car_B.is_Sedan,
car_C.is_Sedan]

count = 0
print("The number of Black cars in inventory is:")
for color in color_list:
    if color == "Black":
        count = count + 1
print(count)
print("The colors of those cars are:", color_list)
```

```
print("Are those cars a Sedan, True or False?",
door_list)
```

Result: *The number of Black cars in inventory is:*

2

The colors of those cars are: ['Black', 'White', 'Black']

Are those cars a Sedan, True or False? [True, True, False]

To count the number of Black car objects, you created a for loop with an embedded if statement. This is cool! You were able to use the Car class in your primary car program. The one thing you forgot to add is what happens when a Black car is not in inventory. Let's make all your car objects "White" and add a clause to says that no cars are **"Black"**:

```
from car_class import Car

car_A = Car("2021", "Tesla", "Model S", "White",
True)
car_B = Car("2021", "Mazda", "CX-5", "White", True)
car_C = Car("2014", "Cadillac", "CTS", "White",
False)

color_list = [car_A.color, car_B.color, car_C.color]
door_list = [car_A.is_Sedan, car_B.is_Sedan,
car_C.is_Sedan]

count = 0
print("The number of Black cars in inventory is:")
for color in color_list:
    if color == "Black":
        count = count + 1
print(count)
if count == 0:
```

```
    print("There are no Black color cars available.")

print("The colors of those cars are:", color_list)
print("Are those cars a Sedan, True or False?",
door_list)
```

Result:

The number of Black cars in inventory is:

0

The colors of those cars are: ['White', 'White', 'White']

Are those cars a Sedan, True or False? [True, True, False]

Little did you know that you would be writing an entire program using classes and objects. You just became an object-oriented programmer in less than 14 days. There's more to this, but you're getting the idea of how this works.

Methods

I went over the data attributes of a class that define what an object is. The second part of a class is its methods. Methods are functions for an object. Each object created can use functions that have been made within its class. Methods provide objects with functions to use.

If you return to the previous example, you created the car Class with data attributes. One of those attributes was an is_Sedan attribute. You will build upon that attribute by giving an object the ability not to be a Sedan using a door change method. Let's go back to your **car_class.py** file. Update the code with the **door_change** method below:

```
class Car:
    def __init__(self, year, make, model, color,
    is_Sedan):
        self.year = year
        self.make = make
        self.model = model
        self.color = color
        self.is_Sedan = is_Sedan

    def door_change(self):
        self.is_Sedan = False
```

The **door_change** method is defined. There are no other parameters associated with it, so the only thing required is self. Self gives us access to **self.is_Sedan** and allows us to assign a variable to it. If this **door_change** method is executed, the object's is_Sedan attribute is automatically set to False despite previous declarations. This means that you can change any car object to a two-door by using the **door_change** method. Copy and paste the code below into another python file called **door_Method.py**:

```
from car_class import Car

car_A = Car("2021", "Tesla", "Model S", "White",
True)

print(car_A.make, car_A.model, "is currently a Sedan
right?")
print(car_A.is_Sedan)
print("Let's use the Door Change method we created
to see if it's still true.")
car_A.door_change()
```

```
print("The", car_A.make, car_A.model, "is still a
Sedan, right?")
print(car_A.is_Sedan)
```

Result: *Tesla Model S is currently a Sedan right?*

True

Let's use the Door Change method we created to see if it's still true?

The Tesla Model S is still a Sedan, right?

False

If you remember, your **car_A** is a 2021 Tesla Model S Sedan. You ran the **door_change** method above, which changed it, so it was no longer a Sedan. This is just one example of how objects can use methods. Methods give objects functions that act as behaviors.

* * *

Q&A Review

1.You can think of a class, as a *blueprint*

1. True
2. False

2.An object is an instance of a class

1. True

2. False

* * *

Day 12 Challenges

- Kilobyte Challenge: Create a Book class and create a Book object.

13

Day 13: Requests Library

Reading Time: 7 minutes

The requests library allows you to send and receive data to and from web servers. By now, you should be getting more familiar with installing third-party libraries. You will use pip to install the requests library, as you have done with other libraries previously.

The requests library allows you to send HyperText Transfer Protocol (HTTP) data. You probably recognize HTTP. Where have you seen it before? You have seen it when you type in a Uniform Resource Locator (URL) in your web browser. For example (https://nerdchallenges.com), the HTTP protocol allows web browsers and other applications to send data back and forth over the internet. HTTP is a protocol that defines how web servers and browsers send data back and forth.

Before installing the requests library and learning how to use it, you will first learn about HTTP Methods and HTTP Response codes.

HTTP Methods

HTTP Methods allow you to send and receive data from a web server. There are four primary methods that you will be using:

- GET
- POST
- PUT
- DELETE

The GET method allows you to request data from a web server, so a GET request is sent whenever you want data from a web server or website. For example, when you go to https://nerdchallenges.com, your web browser will send a GET request to the Nerd Challenges web server.

The POST method allows you to create a new record on a server. In other words, this is for sending data to a web server. It is a way to save data on the webserver. For example, you are ordering a pizza from your favorite pizza restaurant. You go to their website, add your pizza to the cart, and checkout. When you checkout, it sends a POST request to the server to create a *new* order. The pizza restaurant's server will add the order to their database, and they'll begin making your pizza. The POST request is for adding brand-new data to a web server.

The PUT method is similar to the POST method because it adds data to a server. However, it is used when you *overwrite* data with new values. For example, pretend you have a nail salon. In your web application, you have a list of appointments for your customers for the day. Jane, your customer, has an appointment for Tuesday at 4:00 PM, and Jane updates the appointment to Thursday at 5:00 PM. The previous scenario would be a PUT request because Jane already had a record created for an appointment and simply wanted to overwrite that data with new data.

The DELETE method removes data from the webserver or application. Continuing with the nail salon example, Jane now wants to remove her appointment entirely. Jane can send a DELETE request to the webserver to remove her appointment.

There are many more HTTP Methods, but these are the primary four used in almost every application. To see a list of all the HTTP methods, you can visit this website: https://developer.mozilla.org/en-US/docs/Web/HTTP/Methods

HTTP Response Status Codes

An HTTP response status codes tell you whether or not the webserver or application has successfully processed an HTTP request. There are many status codes, but I want to cover three primary ones that you will likely see when you create your applications.

- 200
- 201
- 404

200 is OK. This means that the web server has received the data you have sent or sent data that you asked for, depending on the type of request sent. In essence, it means that everything worked successfully.

201 is Created. This is typically sent after you send data to a server with perhaps a POST or PUT request. It tells you that the record or data has been created and processed.

404 is Not Found. You've likely seen this one when you're browsing a website or an application. This typically means the website owner has moved the page to another place or taken down the page altogether.

Installing requests

To install the requests, run the following command:

```
pip install requests
```

Congratulations, you have now installed requests!

Sending your first GET request

You will import the library and call a '.get' function to send your first GET request. You'll be getting data from https://swapi.dev, a fantastic website with some Star Wars data that you can explore!

```
# Example 1 - Sending your First GET Request
import requests

response =
requests.get('https://swapi.dev/api/people/1/')
print(f"Status Code = {response.status_code}")
print(f"Response from Website = {response.text}")
```

Terminal Output

Status Code = 200

Response from Website =

```
{"name":"Luke Skywalker","height":"172",
...truncated for simplicity
```

Awesome! You just sent your first GET request, and as you can see, you got back data about Luke Skywalker. That's cool! You can see all the properties of Luke Skywalker, such as his eye_color, skin_color, mass, etc. One of the things you may notice is that the response from the website contains all of this text as a string. What if you just wanted to grab a specific property? How could you do this?

You could parse out the data you are looking for using something like 'regex'. This would allow you to search through the data, but this would get very complicated very quickly. Instead, there is an easier way. You could use something called JavaScript Object Notation (JSON). JSON is a way to format data. Keys and values organize it. For example, in the API call you just made, the response from the server is JSON data, where the properties are keys; these would be: name, height, mass, hair_color, etc.

You can use the .json() function to get just the name, height, birth_year, and gender of Luke Skywalker. Here is how to do it

```
# Example 1 - Sending your First GET Request, again
import requests

response =
requests.get('https://swapi.dev/api/people/1/')
print(f"Status Code = {response.status_code}")
print(f"Response from Website = {response.text}")

data = response.json()
```

```
print(type(data))
print(data['name'])
print(data['height'])
print(data['birth_year'])
print(data['gender'])
```

Terminal Output

Status Code = 200

Response from Website =

```
{"name":"Luke Skywalker","height":"172",
...truncated for simplicity
```

You have gotten Luke Skywalker's information and printed the fundamental properties you wanted. You'll also notice that the .json() function returns a Python dictionary. You can easily access the data by typing data['property'].

Sending your first POST request

Recall that POST requests allows you to create new data on a server, let's try and create a new character on the swapi.dev website.

```
# Example 2 - Sending a Post Request? Oh no.
data = {'name': 'John Smith'}
response =
requests.post('https://swapi.dev/api/people/1', data)
print(response.status_code)
print(response.text)
```

Terminal output

405

{"detail":"Method 'POST' not allowed."}

Oh no! You got a 405 error, which, as you can see, means that the web server has refused the request, and the POST method is not allowed on the swapi.dev website. Not all websites and applications accept data from end-users, which is an example of that. Fortunately, there are countless Application Programming Interfaces (APIs) on the internet that you can use. Let's try another one.

Sending your first GET request, again

This time you are sending a GET request to typicode.com, which has some fake blog data we can use. Let's send the request.

```
# Example 3 - Finding a better API :) Sending a GET
request again
# https://jsonplaceholder.typicode.com/posts

response =
requests.get('https://jsonplaceholder.typicode.com/
posts/1')

print(f"Status Code = {response.status_code}")
print(f"Response from Website = {response.text}")
```

Terminal output

Status Code = 200

Response from Website = {

"userId": 1,

"id": 1,

"title": "sunt aut facere repellat provident occaecati excepturi optio reprehenderit",

"body": "quia et suscipit\nsuscipit recusandae consequuntur expedita et cum\nreprehenderit molestiae ut ut quas totam\nnostrum rerum est autem sunt rem eveniet architecto"

}

Again, we could use .json() here to parse out specific data you are interested in, just like you did for the Star Wars API.

Sending your first POST request, for real this time

You're creating a fake blog post with a title and a body and sending it to the server, give it a try.

```
# # Example 4 - Sending a POST again
data = {
 "title": "Python Requests Library is awesome!",
 "body": "Pretty cool yea? :D"
}
response =
requests.post('https://jsonplaceholder.typicode.com/
posts/', data)
print(response.status_code)
print(response.text)
```

Terminal output

201

{

"title": "Python Requests Library is awesome!",

"body": "Pretty cool yea? :D",

"id": 101

}

Notice here that you get a status code of 201, which means the server has added the record. You can see that the server assigned the blog an id number of 101. Let's delete the record now.

DELETE your blog post

This is pretty straightforward, call the .delete() function and pass in the correct URL.

```
# # Example 5 - Sending a DELETE Request
response =
requests.delete('https://jsonplaceholder.typicode.com/
posts/101')
print(response.status_code)
print(response.text)
```

Terminal output

200

{}

You get back a 200, with no data in the text field. The server has successfully deleted the record.

Recap Day 13

In today's lesson, you learned about HTTP methods, HTTP status codes, how to send and receive data with the *requests* library, and how to access APIs. Sending data over the internet using HTTP is a critical skill that every developer should know. Below are two quick Nerd Challenges that you should be able to solve by now. Try and implement them on your own!

* * *

Q&A Review

1.The requests library is built into Python

1. True
2. False

2.The requests library allows you to send a HTTP request

1. True
2. False

* * *

Day 13 Challenges

- Kilobyte Challenge: How could you get additional data about other people besides Luke Skywalker?
- Kilobyte Challenge: If you wanted to get the names of all the Star Wars characters available via swapi.dev, how can you accomplish that?

14

Day 14: Nerd Challenge 3 - Advanced GUI Calculator

Reading Time: 6 minutes

We will go through a step-by-step challenge of creating a calculator with an actual GUI. You made a calculator without any graphical user interface in the first challenge. Everything from the program printed on the terminal screen; while it worked similar to any calculator, it isn't ideal, especially for repeated use as a tool to solve problems. Having a GUI would have significantly increased the user interaction and experience. For this reason, we will be creating a Calculator GUI using the PySimpleGUI package.

Like the last challenge, the Calculator needs to do simple math operations such as add, subtract, multiply, and divide. The only difference is that the user needs a GUI to conduct these math operations. It would be best if you displayed any calculations on the GUI.

It's go time. Are you going to proceed on with being a Python programmer? If so, then say to yourself, **Challenge Accepted!**

When you are all done, your Calculator will look like this:

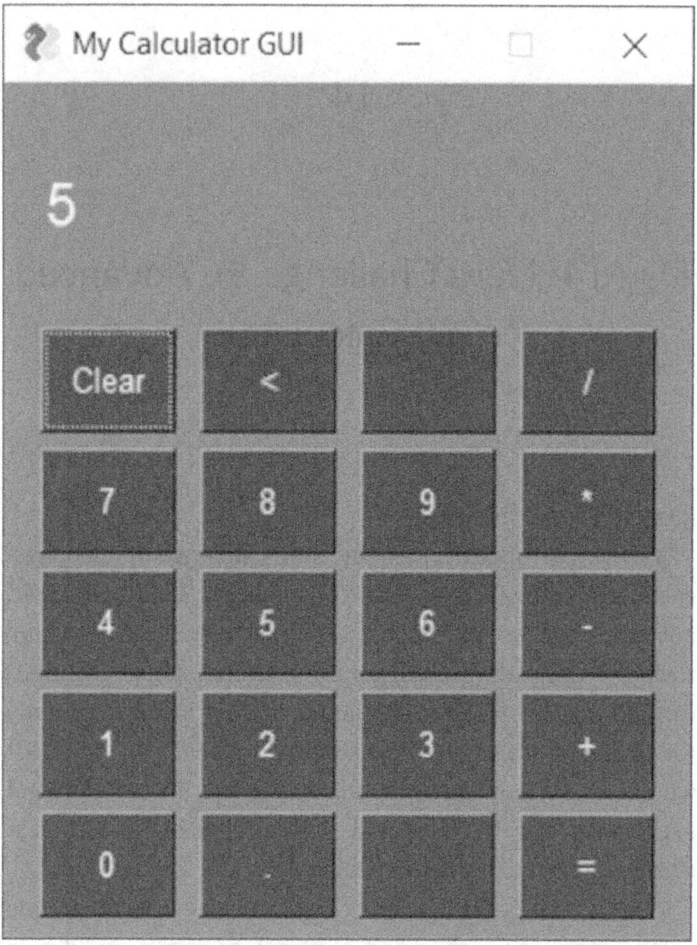

Create a new project called **calculatorGUI**.

When designing a GUI, we must keep in mind the three design elements: window, layout, and event loop. You display the application content on the window. The layout contains the user interface and outputs of the program. The event loop handles what happens when the user interacts with form elements, such as buttons.

Import the module

We first need to import our PySimpleGUI program as we have done in the past. This time we will import the entire module with the syntax **from PySimpleGUI import *** .

Program Greeting

When I create applications, I like to greet a user. This is a calculator, so this may or may not be relevant, but let's do it to get in the mindset of welcoming our users. Use the popup_get_text() for any inputs needed for the greeting, such as name or user. Then use the **popup()** function to print out the greeting. Your greeting could look something similar to this:

```
# Program Greeting
user_name = popup_get_text("Please enter your name
to proceed:")
popup("Hey " + user_name.upper() +", welcome to
Calculator GUI!")
```

Result:

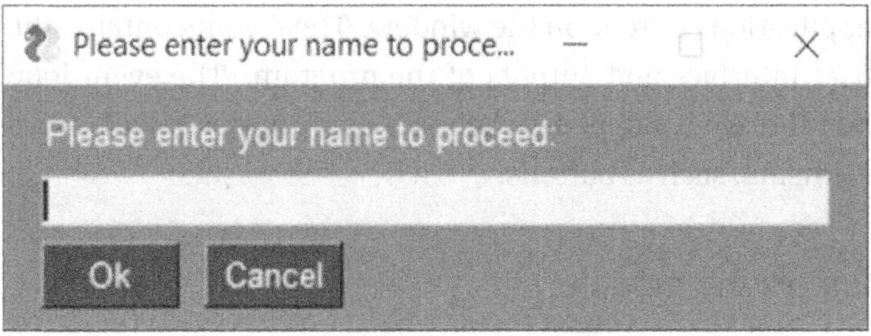

Design the layout

The layout section in PySimpleGUI allows us to output a visual representation of our code in graphics. You have everything from text, buttons, and user input fields to formatting within that layout. PySimpleGUI makes it easy to create layouts without using much code at all.

The entire layout of our calculator program can be represented as a list of lists. The layout is organized into columns and rows with its elements. The first row of the layout will display user inputs and calculated outputs. These values will be displayed on the top row of the layout. This row, the first list, has text formatting attributes that indicate how the values will be outputted. We chose to change the font size, type, and color in our code. The additional attribute needed is the key which is what is outputted based on the event loop section that we will be discussed later:

```
layout = [[Txt('' * 10)], # adds 10 spaces to skip a
line to center the output
          [Text('', size=(17, 1), font=('Arial', 18),
          text_color='white', key='input')],
```

```
        [Txt('' * 10)], # adds 10 spaces to skip a
        line to center the output
```

The last five rows of the layout are the actual buttons. Within the Calculator's buttons, you need numbers (0-9), a decimal, all the arithmetic operators, and the ability to delete and clear. All of these are represented by buttons in the form of lists. Let's add buttons to our previous layout code:

```
layout = [[Txt('' * 10)], # adds 10 spaces to skip a
line to center the output
         [Text('', size=(17, 1), font=('Arial', 18),
         text_color='white', key='input')],
         [Txt('' * 10)], # adds 10 spaces to skip a
         line to center the output
[ReadFormButton('Clear'), ReadFormButton('<'),
SimpleButton(''), ReadFormButton('/')],
[ReadFormButton('7'), ReadFormButton('8'),
ReadFormButton('9'), ReadFormButton('*')],
[ReadFormButton('4'), ReadFormButton('5'),
ReadFormButton('6'), ReadFormButton('-')],
[ReadFormButton('1'), ReadFormButton('2'),
ReadFormButton('3'), ReadFormButton('+')],
[ReadFormButton('0'), ReadFormButton('.'),
SimpleButton(''), ReadFormButton('=')],
         ]
```

Window

The next thing we need to set up is the actual application window. Within this window, we will have a layout. In this section, we are provided with the ability to change the attributes of the layout. Here we designate the width/height in characters of each of the buttons listed in the layout. In this case, we are using size (5,2) for element size. This means five characters wide and two characters high:

```
# Window/From Section: Setup GUI and size buttons
form = FlexForm('My Calculator GUI',
default_button_element_size=(6, 2),
auto_size_buttons=False, grab_anywhere=False)
form.Layout(layout) # chooses the above layout
```

Event Loop

The event loop is where all the operations are conducted. The while loop is what opens the GUI window and keeps it open until the user hits the top right exit button to close the program. The button values are being read in as each button is selected with the following code:

```
# Return Value
Return = ''

# Event Loop
while True:
   # Button Values
   button, value = form.Read() # reads in the values
   from the form when a button is clicked
```

There are some special things that must be set up for our Calculator. For example, if the Clear, < (backspace), X (exit button), or a value over 17 digits is received. Seventeen digits are our limit because our layout only supports 17 characters wide. That can be represented in an if-else statement:

```
# Check Selected Buttons
if button == 'Clear':
   Return = ''
form.FindElement('input').Update(Return) # updates
input key with return value
elif button == '<':
   Return = Return[:-1]
form.FindElement('input').Update(Return)
elif len(Return) == 17: # if return value is over 17
digits no action taken
   pass
# Quit Program
elif button == 'Quit' or button == None:
   break
```

Now that we have addressed those particular cases, it's time to perform the calculator operations. The code below takes in the values from each button pressed and saves it as an expression. That expression is then evaluated so we can receive our final answer. That last answer is then displayed in our key, or output display:

```
# Evaluates the return values in the expression to
get an answer
elif button == '=':
Answer = eval(Return)
Answer = str(round(float(Answer), 3))
form.FindElement('input').Update(Answer)
Return = Answer
else:
Return += button
form.FindElement('input').Update(Return)
```

That last section was all we needed to complete our Calculator GUI program. The next step now is to run our **calculator_GUI.py**. Let's take a look at all three of our sections to see what the entire program looks like:

```
# Import Module
from PySimpleGUI import *

# Program Greeting
user_name = popup_get_text("Please enter your name
to proceed:")
popup("Hey " + user_name.upper() +", welcome to
Calculator GUI!")

# GUI Layout
layout = [[Txt('' * 10)], # adds 10 spaces to skip a
line to center the output
         [Text('', size=(17, 1), font=('Arial', 18),
         text_color='white', key='input')],
         [Txt('' * 10)], # adds 10 spaces to skip a
         line to center the output
[ReadFormButton('Clear'), ReadFormButton('<'),
SimpleButton(''), ReadFormButton('/')],
[ReadFormButton('7'), ReadFormButton('8'),
ReadFormButton('9'), ReadFormButton('*')],
[ReadFormButton('4'), ReadFormButton('5'),
ReadFormButton('6'), ReadFormButton('-')],
[ReadFormButton('1'), ReadFormButton('2'),
ReadFormButton('3'), ReadFormButton('+')],
[ReadFormButton('0'), ReadFormButton('.'),
SimpleButton(''), ReadFormButton('=')],
         ]

# Window/From Section: Setup GUI and size buttons
form = FlexForm('My Calculator GUI',
default_button_element_size=(6, 2),
```

```
auto_size_buttons=False, grab_anywhere=False)
form.Layout(layout)# chooses the above layout

# Return Value
Return = ''

# Event Loop
while True:
   # Button Values
   button, value = form.Read() # reads in the values
   from the form when a button is clicked

   # Check Selected Buttons
   if button == 'Clear':
       Return = ''
form.FindElement('input').Update(Return) # updates
input key with return value
   elif button == '<':
       Return = Return[:-1]
form.FindElement('input').Update(Return)
   elif len(Return) == 17: # If return value is over
   17 digits no action taken
       pass
   # Quit Program
   elif button == 'Quit' or button == None:
       break

   # Evaluates the return values in the expression
   to get an answer
   elif button == '=':
       Answer = eval(Return)
       Answer = str(round(float(Answer), 3))
form.FindElement('input').Update(Answer)
       Return = Answer
   else:
       Return += button
form.FindElement('input').Update(Return)
```

If you could run your program successfully, you would see the windows below.

Result:

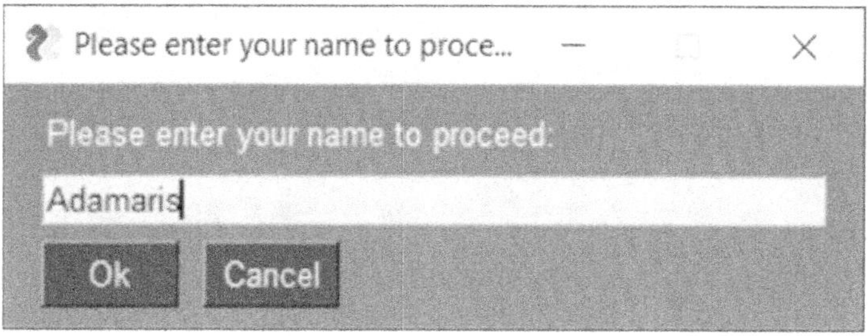

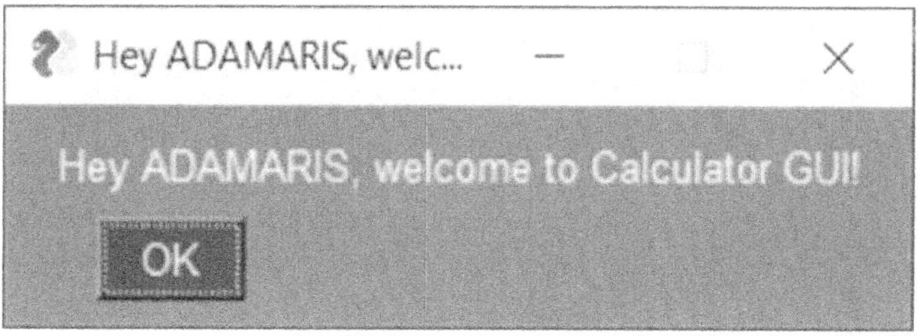

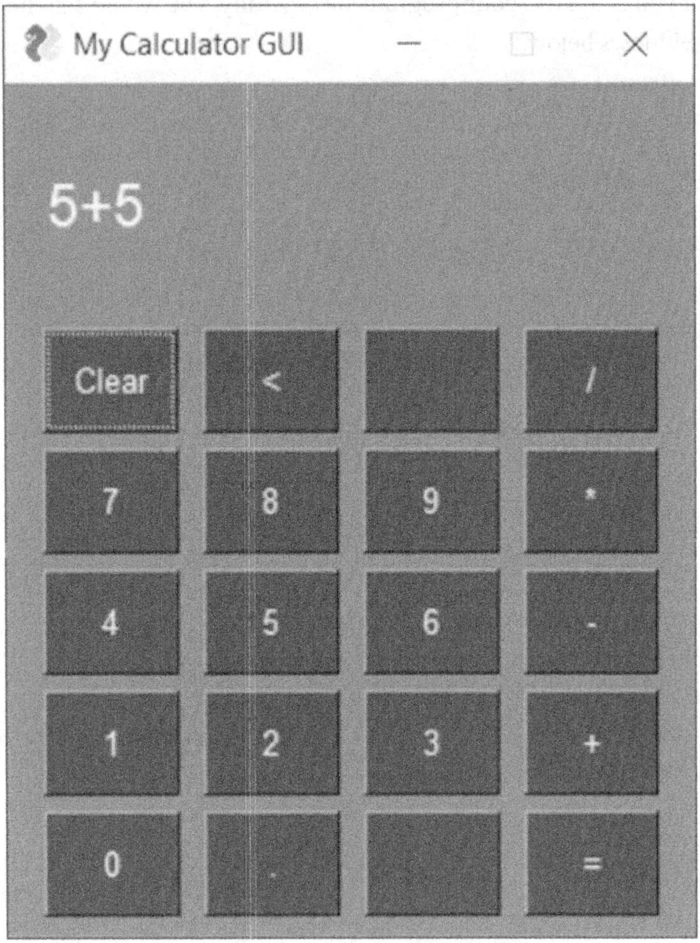
My Calculator GUI
5+5
Clear
<
/
7
8
9
*
4
5
6
-
1
2
3
+
0
.
=

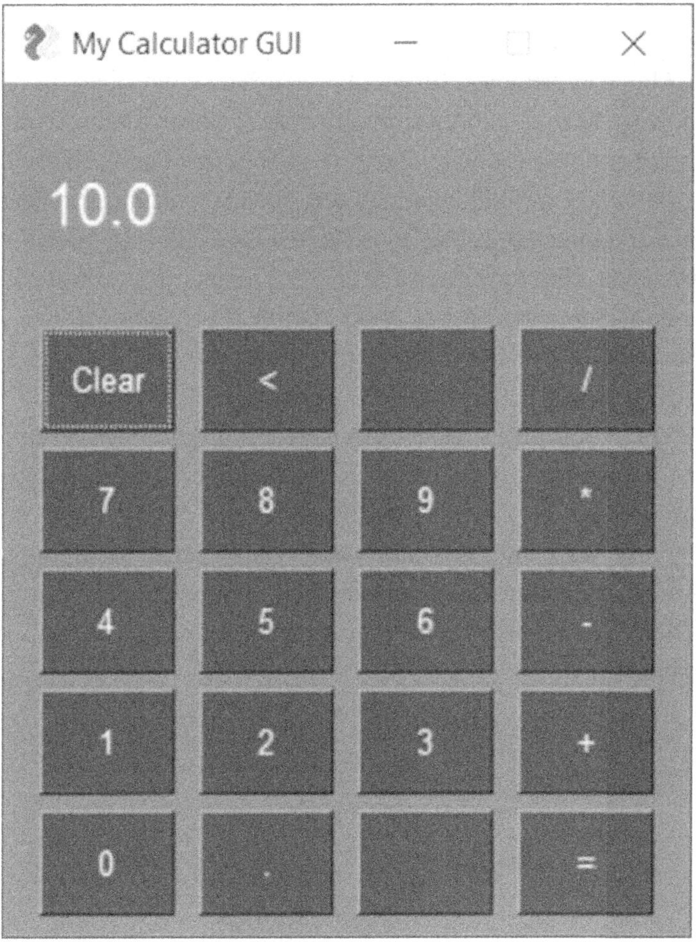

Conclusion

The journey to becoming a python programmer was by no means easy at all. You stuck it out by going through this step-by-step tutorial in learning various programming concepts.

These concepts don't only apply to Python but also to other programming languages.

I hope you enjoyed this book. I hope you enjoyed learning how to become a Python programmer! It was a harrowing journey along the way, but you worked through each step. Casey and I would like to end this book by showing a token of our appreciation. Day 15 is a free lesson from our book on GatsbyJS. This book dives into creating a super-fast website in GatsbyJS in just under 14 days. It's filled with good information, including real-world examples, challenges, and even a supplemental course for even more hands-on practice.

Your Python journey doesn't have to stop here. Python can do many amazing things, from Machine Learning to image processing and DevOps. Also, be on the lookout for our next Python book for some more advanced concepts, including creating applications using object-oriented programming. Thank you again, and we look forward to seeing you again soon.

15

Day 15: Bonus Day - Building a Website with GatsbyJS

Reading Time: 18 minutes

Alright, Nerds! You've made it! You've learned the basics of Python! Woohoo! Python is an excellent language and has many use cases, as you now know. However, on the web today, Javascript is king. In 2022, if I had to pick one framework or front-end stack to learn, it would most certainly be GatsbyJS. Without getting into all of the specifics, here's why:

1. It's super **fast** - primarily because it generates static HTML files to display your sites, which means that it undergoes a build process. You'll learn more about this later. For now, just know that it's super fast.
2. It **scales** - This is really important because since it's easy to scale, you don't have to worry about infrastructure like other web technology stacks (C#, etc.)
3. Follows best practices, especially with **accessibility** and **image optimization.** One of the most fantastic features of GatsbyJS is that many best practices like accessibility and

image optimization are built right into the framework. This doesn't mean you'll never have to do image optimization or that everything is 100% accessible; however, it saves you a considerable amount of work initially.

Those are just some of the benefits. By no means is it an exhaustive list. Today's content is an excerpt from our GatsbyJS 101 book and shows you how to install your IDE, Node.js, and load your first website with GatsbyJS on the Gatsby Cloud. Enough background. Let's dive in, Nerds!

Setting up your environment

As with all programming languages and frameworks, the first thing you must do is set up all of the required tooling. To build a GatsbyJS website, we'll need to install the following tools:

1. Visual Studio Code (https://code.visualstudio.com) - The go-to Node.js editor. As always, though, you're free to choose any IDE you'd like, yes, even notepad and vim.
2. Curl (https://curl.se) - A handy command-line tool that allows you to transfer HTTP/HTTPS data.
3. Homebrew (**macOS only**) - A handy tool that makes installing developer tools super simple on mac.
4. Git CLI (https://git-scm.com) - A free and open-source distributed version control system that you can use to keep track of your source code. You'll need this in conjunction with a GitHub account.
5. Node.js (https://nodejs.org/en/) - A JavaScript runtime that is built on Chrome's V8 JavaScript engine, it allows you to run JavaScript locally on your machine

6. Gatsby CLI (https://www.npmjs.com/package/gatsby-cli) - A command-line interface that you can use to quickly start your Gatsby site.

In this section, you will also find OS-specific instructions for three different Operating Systems:

- Windows 10
- macOS
- Linux (Ubuntu)

You can skip to the respective section for whichever Operating System you are currently using. If you use a different operating system, the instructions should be somewhat similar. For example, if you're using Linux Mint, you can probably follow along via the Ubuntu section. With that being said, all of these tools are well supported and have excellent documentation. If you get stuck on a particular tool, go to the website for that tool and do some searching. The odds are you'll find what you're looking for to resolve the issue.

Installing VS Code

Visual Studio Code is a lightweight, excellent source code editor. What I love most about it is it is cross platform. This means it works on macOS, Windows, and Linux. As a developer, I work on different operating systems depending on the work being done, and it's helpful to keep the IDE experience consistent across other operating systems. Also, VSCode has a rich Plugin ecosystem, which is very powerful. We'll explore some of the better GatsbyJS plugins later.

Windows Setup

Head to https://code.visualstudio.com/download and click download for Windows.

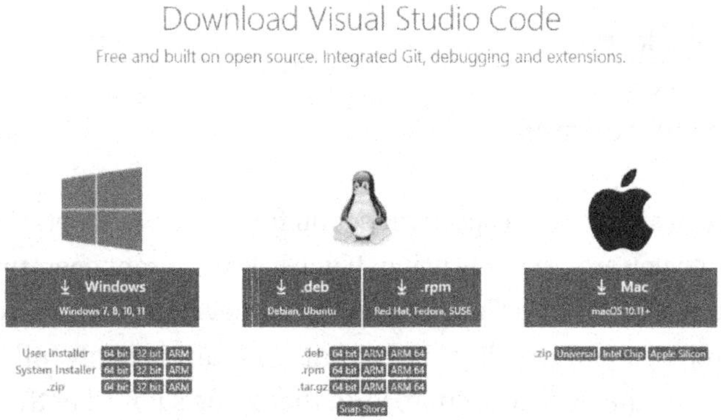

I'll be using the 64-bit User Installer; if you'd like to install it for all users on your system, then use the System Installer. Once the file is downloaded, go ahead and install it. You can use all of the default options. Feel free to change any options that meet your needs.

MacOS Setup

Head to https://code.visualstudio.com/download and click download for Mac. In my case, this is the Apple Silicon; but if you're on the Intel Chipset, then choose that one.

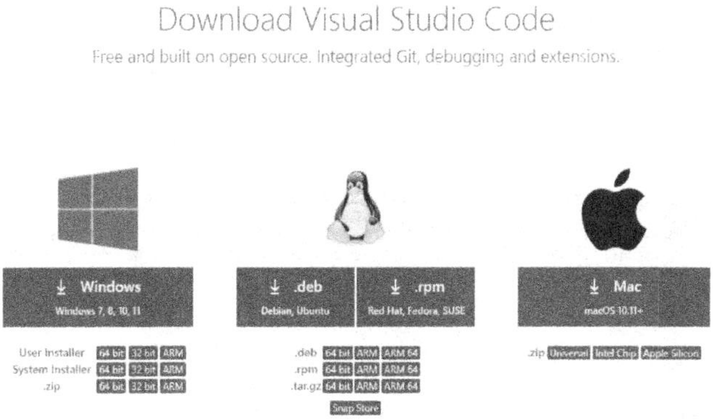

Linux Setup

To install Visual Studio Code on Ubuntu 20.04.2 LTS, you can do it via the Ubuntu Snap Store (https://snapcraft.io/code) or via the command line. To install it graphically, open up the Snap store and search for Visual Studio Code, click Install.

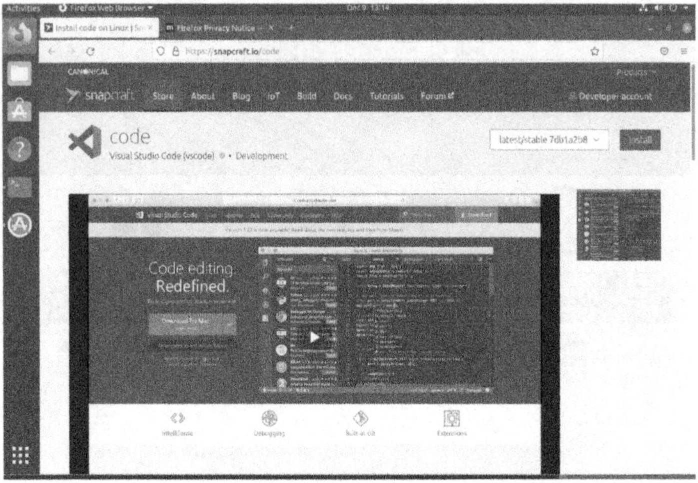

As a developer, though, you should be comfortable operating the terminal. To do this, open up the **terminal** application and enter the command.

```
sudo snap install code --classic
```

You should now be able to launch Visual Studio Code by going to your applications and clicking on Visual Studio Code! Go ahead and do this to make sure everything is set up properly. Alternatively, you can open this up through the **terminal** by typing this command

```
code
```

You should now see Visual Studio Code open! Congrats ⊠!

Installing Curl

cURL is a command-line tool that allows you to transfer HTTP and HTTPS (it also supports other protocols) data via the command line. It is free and open-source.

Windows Setup

Fortunately, as of Windows 10 (and presumably Windows 11), curl is now installed by default. To verify this, open up the Command Prompt and type the following

```
curl --version
```

You should see something similar to the following.

```
Terminal Output
C:\Users\casey>curl --version
curl 7.55.1 (Windows) libcurl/7.55.1 WinSSL
Release-Date: 2017-11-14, security patched:
2019-11-05
Protocols: dict file ftp ftps http https imap imaps
pop3 pop3s smtp smtps telnet tftp
Features: AsynchDNS IPv6 Largefile SSPI Kerberos
SPNEGO NTLM SSL
```

If, for some reason, curl isn't installed, you can download it here: https://curl.se/windows/

macOS Setup

Fortunately, on macOS, curl comes installed by default, but let's verify open up a terminal window and type the following:

```
curl --version
```

You should see something similar to the following:

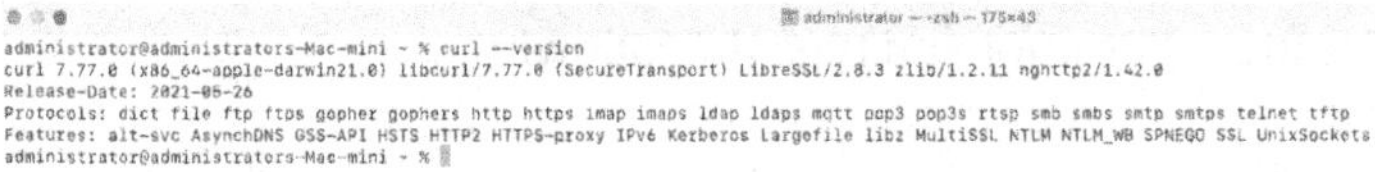

Linux Setup

This updates the package indexes so that you can download the latest files.

```
sudo apt-get update
```

Next, install the curl package.

```
sudo apt-get install curl
```

To verify that curl has been installed, run this command.

```
curl --version
```

You should see:

```
Terminal Output
curl 7.68.0 (x86_64-pc-linux-gnu)
```

Installing git (and Homebrew if MacOS)

You will be utilizing a tool called git, an open-source distributed version control system that allows you to keep track of source code changes over time. Huh? It will help track changes to your code, so if you need to go back in time, you can pretty easily!

Windows Setup

Go to the following URL: https://git-scm.com/downloads and click on Download for Windows. When installing it, you can accept all of the defaults. Once it's installed, open up Command Prompt and type in the following:

```
git --version
```

You should see something similar to the following

```
Terminal output
git version 2.34.1.windows.1
```

MacOS Setup

Homebrew is a developer tool that allows you to easily install packages and software tools on macOS and Linux. Installing it is pretty simple, open up a terminal and run the following command

```
/bin/bash -c \
"$(curl -fsSL \
https://raw.githubusercontent.com/Homebrew/install/
HEAD/install.sh)"
```

Then simply follow the instructions and accept the defaults. You should see something similar to the following output. Note: It may prompt you for your admin password. This is fine.

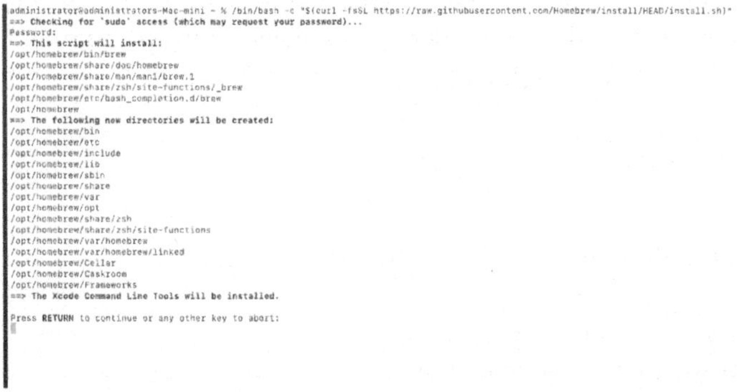

Press RETURN and let it install everything it needs. It will probably take several minutes for Homebrew to download and install everything it needs, so grab some brew ☕

Once Homebrew is installed, you should see a next steps section with two commands. Copy both of those commands and run them. It should look similar to the below but not exactly! So don't copy these. Copy the ones in your terminal.

```
echo 'eval "$(/opt/homebrew/bin/brew shellenv)"' >>
/Users/administrator/.zprofile
eval "$(/opt/homebrew/bin/brew shellenv)"
```

Awesome you'll use brew to install git! Install git by running this command:

```
brew install git
```

After it's done just type the following:

```
git --version
```

If it reported a version, then you're good to go! All done! Life is so easy with Homebrew!

Linux Setup

Install git by running this command:

```
sudo apt-get install git
```

To verify that git has been installed, run this command:

```
git --version
```

It should print out a version, such as git version 2.25.1.

Installing Node.js

Node.js allows you to run Javascript locally on your machine. Since you're using Javascript to build your website, you need to be able to run and test things outside of your browser. React and GatsbyJS utilize Node.js to create artifacts that you will later deploy to a hosting provider.

Additionally (on macOS and Linux), you can use **Node Version Manager (NVM)** to switch between different versions of Node.js easily. While you may not necessarily use this feature, it is a handy tool if you ever need to swap versions of Node.js for a different project. To learn more about NVM, you can look at the GitHub Repo located here https://github.com/nvm-sh/nvm

Windows Setup

Head over to https://nodejs.org/en/download/ and download the Windows Installer (.msi) file for the LTS version. At the time of this writing, it is v16.13.1; however, any version after v14 will likely suffice. Once you download it, go ahead and install it and accept all of the default installation options.

After it is installed, open up your Command Prompt and type the following:

```
node --version
```

You should see something similar to the following

```
Terminal output
v16.13.1
```

Sweet! It's installed.

MacOS Setup

Because you installed Homebrew earlier this is a breeze, just run the following command

```
brew install nvm
```

After that, you'll see some next steps that homebrew and nvm ask you to do, which result in creating a directory and updating your configuration files. Be sure to follow those instructions, which should look something like this.

```
==> Caveats
Please note that upstream has asked us to make explicit managing
nvm via Homebrew is unsupported by them and you should check any
problems against the standard nvm install method prior to reporting.

You should create NVM's working directory if it doesn't exist:

  mkdir ~/.nvm

Add the following to ~/.zshrc or your desired shell
configuration file:

  export NVM_DIR="$HOME/.nvm"
  [ -s "/opt/homebrew/opt/nvm/nvm.sh" ] && . "/opt/homebrew/opt/nvm/nvm.sh"  # This loads nvm
  [ -s "/opt/homebrew/opt/nvm/etc/bash_completion.d/nvm" ] && . "/opt/homebrew/opt/nvm/etc/bash_completion.d/nvm"  # This loads nvm bash_completion

You can set $NVM_DIR to any location, but leaving it unchanged from
/opt/homebrew/opt/nvm will destroy any nvm-installed Node installations
upon upgrade/reinstall.
```

If you don't follow those instructions, nvm may not work correctly. Your final configuration file should look something like this:

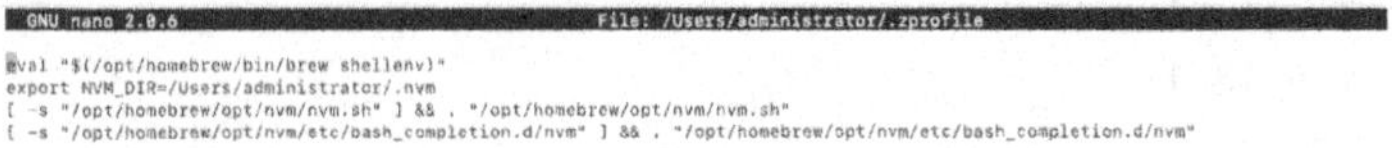

Now that nvm is installed let's verify everything is working as expected by typing

```
nvm --version
```

nvm should report the version number and it should look something like this.

```
Terminal Output
0.39.0
```

Now that we have nvm installed, we can use it to install Node.Js. To do this, type the following command:

```
nvm install --lts
```

This should install the **LTS** version of Node.Js, which stands for Long Term Support. The LTS term means that Node.Js will support that primary LTS version for 3 years. As of this writing, the current LTS version is v16.13.1. I recommend using the LTS version as it is more stable than the latest version of Node.js, which is currently v17.2.0.

However, if you'd like to try out the latest version of Node, you can use nvm to install it and then simply switch between the two. To do this, simply run:

```
nvm install node
```

This will install the latest version of Node.js. You can switch to the latest version by typing:

```
nvm use node
```

To switch back to the LTS version, simply type:

```
nvm use --lts
```

Pretty awesome right?!

Linux Setup

nvm is a bash script. To install it run this command:

```
wget -o- \
https://raw.githubusercontent.com/nvm-sh/nvm/
v0.38.0/install.sh
```

This will download the install.sh script, you can then run it with:

```
bash install.sh
```

Next, close your terminal and re-open the terminal application again. This will refresh your terminal so you can access **nvm**. To check that **nvm** is appropriately installed, type this command:

```
nvm --version
```

nvm should report the version number, and it should look something like this:

```
Terminal Output
0.39.0
```

Now that we have nvm installed, we can use it to install Node.Js. To do this, type the following command:

```
nvm install --lts
```

This should install the **LTS** version of Node.Js, which stands for Long Term Support. The LTS term means that Node.Js will support that primary LTS version for 3 years. As of this writing, the current LTS version is v16.13.1. I recommend using the LTS version as it is more stable than the latest version of Node.js, which is currently v17.2.0.

However, if you'd like to try out the latest version of Node, you can use nvm to install it and then simply switch between the two. To do this, simply run:

```
nvm install node
```

This will install the latest version of Node.js. You can switch to the latest version by typing:

```
nvm use node
```

To switch back to the LTS version, simply type:

```
nvm use --lts
```

Pretty awesome right?!

Installing Gatsby CLI

Installing Gatsby CLI is the same process on Windows, macOS, and Linux. The Gatsby CLI will help you quickly create a new Gatsby website. It makes getting started extremely simple. The tool will set up the entire website package so you can get started building.

You're going to install it as a global npm package since you'll use it across multiple projects. Generally speaking, you should avoid installing packages globally; however, it can make sense in certain situations for some CLI-based tools. In this case, it does. To install it, run:

```
npm install --global gatsby-cli
```

Viola! Gatsby CLI is installed. Make sure it is working by running:

```
gatsby --version
```

You should see it return Gatsby CLI Version: 4.3.0 (or a newer version)

Hello World with GatsbyJS v4

Alright! All your developer tools are installed. Whew, that was a journey! Now it's time to get your first GatsbyJS v4 site up and running in your development environment.

You will use the Gatsby CLI to create the initial boilerplate website. Open up a terminal window and navigate to whichever folder you'd like to make the website in. For example, I put all of my projects in a folder named external-website-projects. So, inside VSCode, open up a terminal, and I can navigate to that directory by typing:

```
cd ~/Documents/external-website-projects/
```

This is my directory, but it doesn't have to be yours. Feel free to create whatever directory structure you like. I'm simply going to refer to this location as the **website project folder** in the future. Once you are where you want to be, create a new Gatsby project by typing in:

```
gatsby new
```

The Gatsby CLI will ask you several questions, which it will use to determine specific build settings. So, you are telling this command-line program, gatsby, to create a new site for you. Answer the questions Gatsby CLI asks as follows:

What would you like to call your site?

```
My First Gatsby Site
```

What would you like to name the folder where your site will be created?

```
my-first-gatsby-site
```

Will you be using a CMS?

```
No (or I'll add it later)
```

Would you like to install a styling system

```
No (or I'll add it later)
```

(Multiple choice) Use arrow keys to move, enter to select, and choose "Done" to confirm your choices

```
Don't select any of the choices, and select Done
```

Shall we do this (Y/n)?

```
Yes
```

Awesome! Our site is now built, time to take a peek and check it out. Navigate to the site directory.

```
cd my-first-gatsby-site
```

Launch your site locally by typing:

```
gatsby develop
```

This command tells Gatsby to start the local development server on port 8000 as well as the GraphiQL tool on the '___*graphql path*', you will learn about GraphiQL in a later chapter. For now, If you open up your web browser to the following address http://localhost:8000/ you should see your first GatsbyJS site!

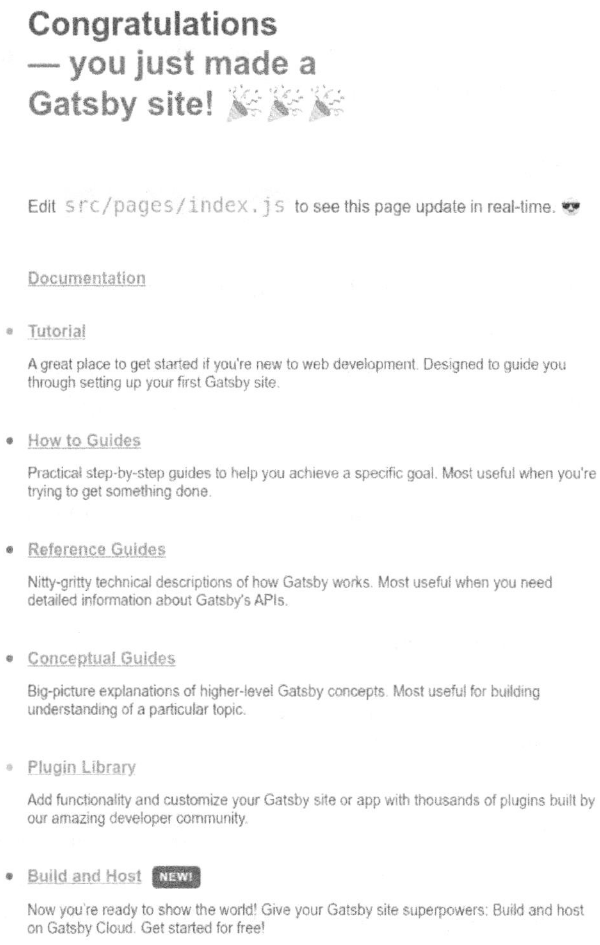

Pretty awesome, right? Now that you see how quickly you can get started with building a website on GatsbyJS, it's time to make some modifications.

Updating the Home page

In the VSCode IDE, open up the src folder, next open the pages folder, inside of there you should see two files:

1. index.js
2. 404.js

index.js is your homepage. This is what will be displayed when you navigate to http://localhost:8000. The 404.js page will be displayed when you go to a link that doesn't exist on your site. Go ahead and open up index.js on line 136. You should see the following HTML code:

```
<span style={headingAccentStyles}>-- you just made a
Gatsby site! </span>
```

Update this line to as follows:

```
<span style={headingAccentStyles}>-- I just made my
FIRST Gatsby site</span>
```

Save the file in your terminal. Notice if you keep your web browser open when you save the file. Gatsby automatically refreshes the page for you to see your changes take effect immediately!

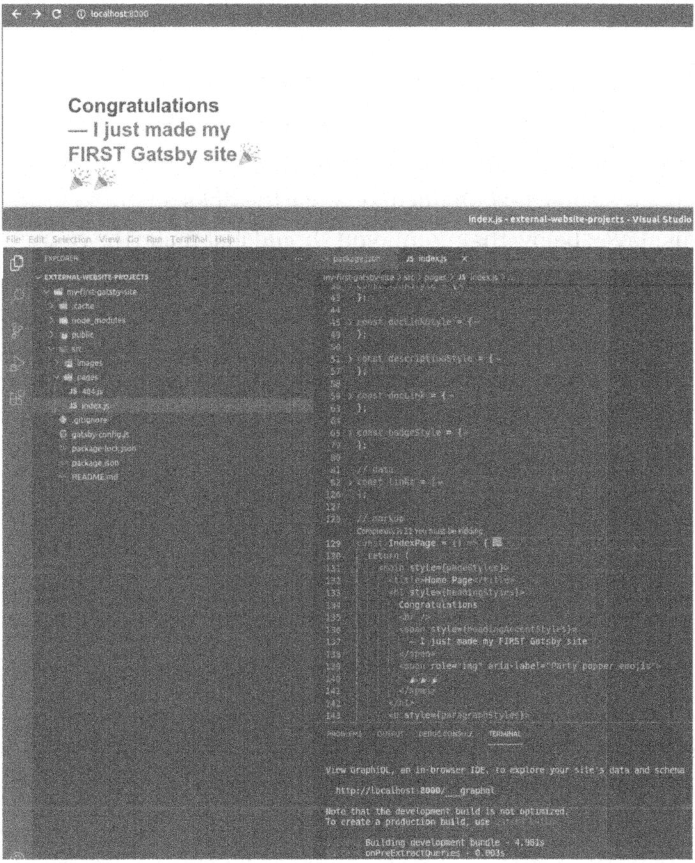

As you can see, it's pretty easy to make a change to a Gatsby site. In fact, if you already know HTML, this was a pretty simple exercise, open up a file and make an HTML change. Since Gatsby uses React, all of the pages will contain some combination of Javascript, HTML, and CSS.

Creating a new link

Now that you've made a text change to the HTML, it's time to create a new link. You would typically go to just below where you made your initial change, right? After all, you just updated some text on line 136. Since the links on the page are displayed below that point, you should just make the change there, correct? Below is what you will find.

```
{links.map((link) => (
          <li key={link.url} style={{
          ...listItemStyles, color: link.color }}>
            <span>
              <a
                style={linkStyle}
                href={`${link.url}...truncated`}
              >
                {link.text}
              </a>
              {link.badge && (
                <span style={badgeStyle}
                aria-label="New Badge">
                  NEW!
                </span>
              )}
              <p style={descriptionStyle}>
                {link.description}
              </p>
            </span>
          </li>
        ))}
```

If you've been using Javascript for some time, you know exactly what to do next, but if **Javascript** or **JSX** is somewhat new to you, you might be a little confused.

- Where are all your HTTP links?
- What do these mean {}?
- What does ‘links.map‘ do in that first line?

You’ll be learning about all of these concepts as you progress through the following chapters. What I want you to take away from this is that with GatsbyJS, you can separate your data from how it is displayed.

Okay, I get it. I think? So how do I make this change?

Scroll up to the top of the file on line 82 you’ll see the following:

```
const links = [...]
```

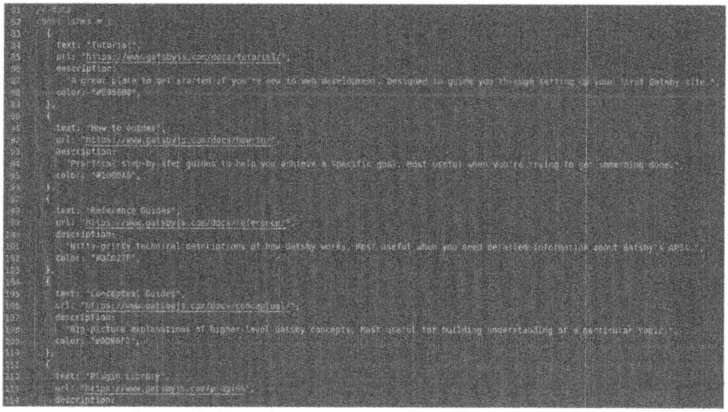

Aha! (Btw, you can shrink the links array by simply moving your mouse over line 82 and clicking the down arrow. VSCode will collapse the links array. This is handy when you're working on code so you can shrink sections you aren't currently modifying)

So here you have the **links** data array, which stores all of the information about a link. It has four required properties: text, URL, description, color, and an optional fifth one, badge. So, go ahead and create a new link by copying everything in between the {} brackets (Don't forget the comma!). Make the change as follows

```
const links = [
  {
    text: "Nerd Challenges",
    url: "https://nerdchallenges.com",
    description:
      "A great place to challenge yourself to learn
      new skills, like GatsbyJS and Python!",
    color: "#E90025",
  },
  {
    text: "Tutorial",
    url: "https://www.gatsbyjs.com/docs/tutorial/",
    description:
      "A great place to get started if you're new to
      web development. Designed to guide you through
      setting up your first Gatsby site.",
    color: "#E95800",
  },
  ***truncated***
```

Save the file, and voila! The page now loads your latest link.

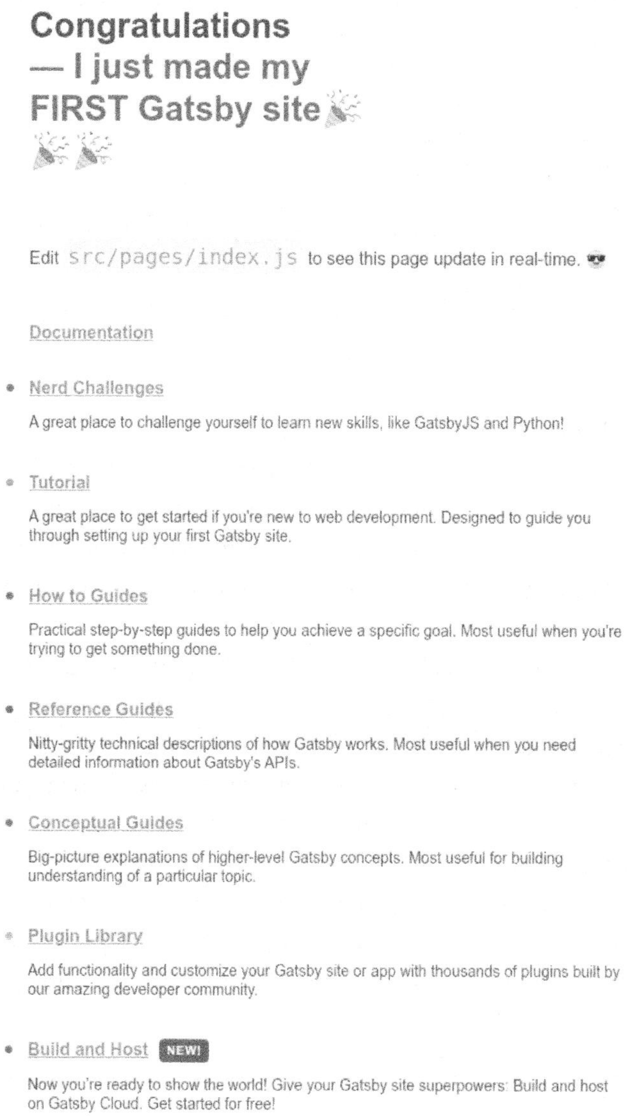
Congratulations — I just made my FIRST Gatsby site
Edit src/pages/index.js to see this page update in real-time.
Documentation
Nerd Challenges
A great place to challenge yourself to learn new skills, like GatsbyJS and Python!
Tutorial
A great place to get started if you're new to web development. Designed to guide you through setting up your first Gatsby site.
How to Guides
Practical step-by-step guides to help you achieve a specific goal. Most useful when you're trying to get something done.
Reference Guides
Nitty-gritty technical descriptions of how Gatsby works. Most useful when you need detailed information about Gatsby's APIs.
Conceptual Guides
Big-picture explanations of higher-level Gatsby concepts. Most useful for building understanding of a particular topic.
Plugin Library
Add functionality and customize your Gatsby site or app with thousands of plugins built by our amazing developer community.
Build and Host
NEW!
Now you're ready to show the world! Give your Gatsby site superpowers: Build and host on Gatsby Cloud. Get started for free!

But wait, we're missing that shiny NEW badge?! No problem, add the following to the top entry.

```
{
    text: "Nerd Challenges",
    url: "https://nerdchallenges.com",
    description:
      "A great place to challenge yourself to learn
      new skills, like GatsbyJS and Python!",
    color: "#E90025",
    badge: true,
  },
```

Boom! Do you see the impact of this? Take a step back, so you were able to modify a piece of data, specifically a Javascript array, which then Gatsby used to render the page differently. You didn't have to style anything with CSS or make HTML changes. You just simply changed a piece of data to add a new record.

What's powerful about this is that later you will learn how to query data from an external data source, such as a MySQL database or an external API. Say you wanted to pull a list of stock prices from a data source (API) and display them on your website. You can query the data with Gatsby via GraphQL, and Gatsby will display that data for you. This means that any time you query that data source, it will show on your site if any new data has been added! Once you program everything correctly, it won't require any updates other than just re-running the query to fetch the latest data!

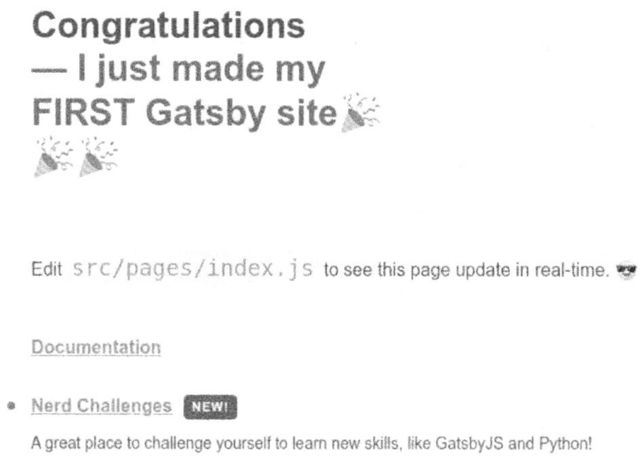

In this chapter, you've learned a few key things

1. How to set up an initial project with the Gatsby CLI. Feel free to create a new project and break stuff! ⊠
2. How to edit text on a Gatsby page
3. How to edit a data array which then subsequently renders different data on a Gatsby page
4. The critical distinction between updating a data object vs. updating HTML tags manually

Tomorrow, you're going to take a step back from Gatsby, and we'll focus on the project, what the goal is we are trying to achieve, and what the requirements are from our customers.

Q&A Review

1.What is GatsbyJS?

1. A React based framework that allows you to build fast, secure, and scalable websites.
2. A static site generator
3. A way to build a website
4. It's that Jay guy right?

2.What is the command to start the local development server with Gatsby CLI?

1. gatsby start
2. gatsby develop
3. gatsby new
4. gatsby stop

3.What is the command to create a web application project with Gatsby CLI?

1. gatsby develop
2. gatsby start
3. gatsby new
4. gatsby stop

Day 15 Challenges

This wouldn't be a Nerd Challenges book without some challenges right! So get after it!

- Kilobyte Challenge: Try and update the Site title on the website you developed to say My First Gatsby Page!.
- Megabyte Challenge: Try and change the font color of the **Congratulations — I just made my FIRST Gatsby site** to #E90025
- Gigabyte Challenge: Try and update the **Documentation** link on the web page
- Terabyte Challenge: Try and deploy this site yourself to Gatsby Cloud!

It's been fun!

Ricardo and I hope you enjoyed this Python book and this little excerpt from our GatsbyJS v4 book. We had a blast writing it and recording the videos, and we hope you learn from it. As always, feel free to reach out to us and tell us what you thought about the book, good or bad! casey@nerdchallenges.com and rico@nerdchallenges.com

Solutions to Q&A

Day 1 - Installing Python Answers

1.What is an IDE?

Integrated Development Environment: Basically a text editor that gives you a bunch of tools to help you write better software.

2.What is the best IDE software?

This is a trick question! It's whatever I like the best, for my personal reasons.

3.What is a variable?

A variable allows you to reference data by specifying a property, you can later reference that property in your application. For example my_name = "Casey" would be a variable.

4.How do you print something to the terminal window in Python 3?

```
print("This doesn't print anything!")
```

5.Why should you use comments?

To provide meaningful context to a python program

6.Can you have multi-line comments?

True

Day 2 - Data Types

1.How do you convert a number to a string, explicitly?

By using the *str()* method

2.Tuples are immutable?

True

Day 3 - Operators

1.What operator is used to multiply two numbers together?

*****

2.There are comparison operators in Python, which allow you to compare integers.

True

Day 4 - User Interaction

1.What Python function is used to prompt the user for input?

input()

2.When a user responds to an input function with a number, the variable is saved as a string.

True

Day 5 - If-Else Statements

1.What type of statement in Python can be used to make a decision?

if-else statements

2.Suppose you have the below code block and it is executed, what will you see on the terminal?

```
rider_age = 13
age_limit = 16
if rider_age >= age_limit:
  print("Congrats! You are old enough to go on this
  ride!")
```

Nothing will be outputted

3.In order to use a 'if' statement, you must also use an else.

False

Day 6 - Try-Exceptions

1.How do you handle errors or exceptions within a Python program?

try-except blocks

2.What is the term used to describe the joining or combining of strings?

Concatenation

3.Finally blocks always run

True

Day 8 - Functions

1.How do you create or define a function in your code?

def

2.How many times are you allowed to use a function in a program?

Unlimited

Day 9 - Modules & Packages

1.What keyword do you use to use a package or module in your Python code

import

2.What is the package installer / manager for Python?

pip

Day 10 - Working with Files

1.Using the open and write function for an existing file:

will overwrite all the contents already in that file

2.The following code block will open a file called nerd_names.txt, read it, and print it to the terminal and then subsequently close the file.

```
nerd_file = open("nerd_names.txt", "r")
print(nerd_file.read())
nerd_file.close()
```

True

Day 11 - Debugging

1.A breakpoint allows you to

allows you to stop your program at a specific line number and then step through subsequent line codes

2.A watcher allows you to watch a variable as your program runs.

True

Day 12 - Classes and Objects

1.You can think of a class, as a *blueprint*

True

2.An object is an instance of a class

True

Day 13 - Requests Library

1.The requests library is built into Python

False

2.The requests library allows you to send a HTTP request

True

Day 15 - Bonus Day - Building a Website with GatsbyJs

About the Author

Ricardo Reid

Ricardo A. Reid is an engineer and military officer with over 12 years of Systems Engineering experience in the aerospace and defense industry. He has worked for Lockheed Martin and Northrop Grumman to design, develop, and integrate weapon systems for the U.S. military. Notable programs include the development of hypersonic missiles and dual-band sensors for 5th generation fighter jets. As a Navy Reservist, he has honorably served his country for 15 years and completed two tours to Afghanistan. He currently serves as a Commanding Officer for a unit that provides depot-level maintenance support to naval surface ships and submarines.

He is an avid learner and always dedicated to continual education and teaching. He is a graduate of the University of Central Florida. He received a B.S. in Electrical Engineering, M.S. in Systems Engineering, and M.S. in Engineering Management. As of December 2015, he is a certified Project Management Professional (PMP) with various IT and cloud certifications. He holds 5 AWS certifications, 3 CompTIA certifications, and previously held a CCNA certification.

Ricardo has a strong passion for teaching technical skills to others, especially military veterans and spouses. His goal is to better educate as many people as possible in engineering and information technology. He is devoted to helping others understand various technical topics with the least effort. He believes everyone is capable of acquiring technical skills. All it takes is a little bit of effort and a will to learn. When it comes to technology, the sky is the limit!

Casey Gerena

Casey Gerena started building software at the young age of eight and has worked for companies such as Amazon, Accenture, and Lockheed Martin in various engineering-related roles. Casey became passionate about software at the young age of seven when he played the video game Doom on his PC. Casey is passionate about teaching others software engineering, the cloud, and serverless technology. He believes serverless is the future of software engineering. Casey holds a BS in Management Information Systems from the University of Central Florida, an MS in logistics and global supply chain management from Embry-Riddle 229 Aeronautical University, and a second MS in Computer Science from the Georgia Institute of Technology. Casey holds several IT certifications, including the Certified Information Systems Security Professional (CISSP) and nine AWS certifications. Casey loves helping others learn about technology, especially when it comes to software engineering. He enjoys weightlifting, playing video games, and spending time with his wife and daughter, Marilyn and Mia.

You can connect with me on:

- https://nerdchallenges.com
- https://twitter.com/cjgerena
- https://twitter.com/rricoreid

Subscribe to my newsletter:

- https://nerdchallenges.com

Made in the USA
Monee, IL
04 November 2026

67382999R10144